Lecture Notes
in Economics and
Mathematical Systems

Managing Editors: M. Beckmann and H. P. Künzi

Mathematical Economics

138

Koji Okuguchi

Expectations and Stability
in Oligopoly Models

Springer-Verlag
Berlin · Heidelberg · New York 1976

Author

Koji Okuguchi
Department of Economics
Tokyo Metropolitan University
1-1-1 Yakumo
Meguro-ku
Tokyo/Japan

Library of Congress Cataloging in Publication Data

Okuguchi, Kōji, 1935-
    Expectations and stability in oligopoly models.

    (Lecture notes in economics and mathematical systems ;
138 ; Mathematical economics)
    Includes bibliographical references.
    1. Oligopolies--Mathematical models. I. Title.
II. Series: Lecture notes in economics and mathematical
systems ; 138.
HD2731.O38   338.5'23   76-30340

AMS Subject Classifications (1970): 90A15, 90D45

ISBN-13: 978-3-540-08056-5   e-ISBN-13: 978-3-642-46347-1
DOI: 10.1007/978-3-642-46347-1

Acknowledgements

I am grateful to Professor Ryuzo Sato of Brown
University for valuable comments and suggestions as
well as encouragement; to an anonymous reviewer for
comments; to Professor Takashi Negishi for his kind-
ness in permitting to use results of a joint work;
to Professor J.W.Friedman for giving me access to
his draft of a forthcoming book Oligopoly Theory; to
Professors J.B.Cruz Jr., L.W.McKenzie, Kotaro Suzumura
and Moriyuki Yoshioka; to Dr.R.D.Theocharis; and finally to the editors of Review of Economic Studies
and Zeitschrift für Nationalökonomie for permission
to use results published in their journals.

TABLE OF CONTENTS

INTRODUCTION 1

CHAPTER 1. EXISTENCE AND STABILITY OF THE COURNOT OLIGOPOLY
            SOLUTION (OR EQUILIBRIUM) 4
    1.1. No Product Differentiation 4
    1.1.1. Existence Proof 4
    1.1.2. Stability Analysis 9
    1.2. Product Differentiation 17
    1.3. Mathematical Appendix 21
    1.3.1. A Theorem on Contraction Mapping 21
    1.3.2. Global Stability of a Discrete System 22

CHAPTER 2. UNIQUENESS OF THE COURNOT OLIGOPOLY SOLUTION 27

CHAPTER 3. ENTRY IN THE COURNOT MODEL: QUASI-COMPETITIVENESS
            VS PERFECT COMPETITION 32
    3.1. Introductory Remarks 32
    3.2. Quasi-Competitiveness 33
    3.3. Convergence to Perfect Competition 36

CHAPTER 4. REVENUE MAXIMIZING DUOPOLY 39
    4.1. Introduction 39
    4.2. Stability Analysis 42

CHAPTER 5. STACKELBERG DUOPOLY MODELS RECONSIDERED 46
    5.1. A Leader-Follower Model 46
    5.2. Resolution of Stackelberg Disequilibrium
          in a Leader-Leader Model 49
    5.2.1. Introductory Remarks 49
    5.2.2. Local Stability 50
    5.2.3. Global Stability 53
    5.2.4. Unknown Market Demand Function and Global Stability. 54

CHAPTER 6. EXTRAPOLATIVE EXPECTATIONS AND STABILITY OF
            OLIGOPOLY EQUILIBRIUM 56
    6.1. Introduction 56
    6.2. Stability under No Product Differentiation 57
    6.3. Product Differentiation and Stability 63

CHAPTER 7. ADAPTIVE EXPECTATIONS AND STABILITY OF
            OLIGOPOLY EQUILIBRIUM 66
    7.1. No Product Differentiation 66
    7.2. Product Differentiation 72
    7.3. Mathematical Appendix 75
    7.3.1. Some Theorems on Matrices with Quasi-Strictly
            Dominant Diagonal Blocks 75
    7.3.2. Some Applications 78

CHAPTER 8. UNKNOWN DEMAND FUNCTION AND STABILITY 82
    8.1. Introduction 82
    8.2. The Cournot Model with Unknown Market
          Demand Function 82
    8.3. Adaptive Expectations and Unknown Demand Function 85

CHAPTER 9. PROBABILITY MODELS 89
    9.1. Probability Models with No Bayesian Learning 89
    9.2. Bayesian Learning in Duopoly Models 95

REFERENCES 99

INTRODUCTION

Ever since A.C.Cournot(1838), economists have been increasingly
interested in oligopoly, a state of industry where firms producing
homogeneous goods or close substitutes are limited in number. The
fewness of firms in oligopoly gives rise to interdependence which they
have to take into account in choosing their optimal output or pricing
policies in each production period. Since each firm's profit is a
function of all firms' outputs in an oligopoly without product differentiation, each firm in choosing its optimal output in any period has
to know beforehand all other rival firms' outputs in the same period.
As this is in general impossible, it has to form some kind of expectation on other firms' most likely outputs. Cournot thought that in
each period each firm assumed that all its rivals' outputs would remain
at the same level as in the preceding period. Needless to say, the
Cournot assumption is too naive to be realistically supported. However,
the Cournot profit maximizing oligopoly model characterized by this
assumption has many important and attractive properties from the view-
point of economic theory and provides a frame of reference for more
realistic theories of oligopoly. In Chapters 1-3, we shall be engaged
in analyzing the Cournot oligopoly model in greater detail from the
viewpoints of existence, stability, uniqueness and quasi-competitiveness of the equilibrium.

It is now widely recognized (See W.J.Baumol(1958, 1959)) that the
oligopolist's objective is revenue maximization under a minimum profit
constraint rather than unconstrained profit maximization. In Chapter
4, the stability of the equilibrium in a revenue maximizing duopoly
model involving the Cournot assumption will be analyzed.

According to a classification by H.von Stackelberg(1934), three
duopoly models can be considered; they are

(a) follower-follower model where both duopolists are followers,

(b) leader-follower model where one firm is a leader and the
other a follower, and

(c) leader-leader model where both duopolists strive for leadership.

The follower-follower model is none other than the Cournot assumption
which is analyzed in Chapters 1-3. In Chapter 5, we shall be concerned
with dynamic analysis of the leader-follower and leader-leader duopoly
models. In contrast to Stackelberg's assertion of inevitability of
economic warfare in the leader-leader model, it will be shown in Chapter
5 that the equilibrium is stable not only locally but also globally

under reasonable conditions provided that each duopolist is assumed to perceive subjectively its rival's marginal cost function and the market demand function.

The Cournot assumption is a special case of more realistic and general types of expectations, extrapolative and adaptive expectations. When expectations are extrapolative and time is discrete, expectation on an economic variable for period t equals its actual value in period t-1 plus the difference between its actual values in period t-1 and t-2 multiplied by a coefficient whose absolute value is less than unity. Under adaptive expectations, expectation for period t is equal to expectation for period t-1 plus the difference between the actual value and expectation both for period t-1 multiplied by a positive coefficient less than or equal to one. When the relevant coefficients in both types of expectations become zero, the Cournot-type assumption is seen to emerge. In Chapters 6 and 7, we shall introduce extrapolative and adaptive expectations into price or quantity adjusting oligopoly models and investigate the stability of the equilibrium in dynamic oligopoly models.

In all of the oligopoly models up to Chapter 7 except for Subsection 5.2.4, it is assumed that all firms are in a position to know correctly the market demand function. More often than not, firms in oligopoly are unlikely to have access to all informations which will enable them to infer correctly the market demand function. In Chapter 8, which is intended to cope with this situation of incomplete information, we shall assume that each firm has its own subjective market demand function. The effects of the Cournot assumption and adaptive expectations on the stability of the equilibrium will be brought to light there.

In general, oligopolists will have to confront uncertainties in three respects;

(a) Firms do not correctly know rivals' cost functions.

(b) As already mentioned, firms do not have perfect information about the market demand function.

(c) When firms have to choose optimal outputs or prices in any period, they can not correctly estimate rivals' outputs or prices in the same period.

Our analyses up to Chapter 8 will be taken as partial answers to problems pertaining to uncertainties facing oligopolists. How partial these answers are may be seen in total lack of intrinsic probabilistic considerations. When firms do not exactly know rivals' outputs (or prices) and/or the true market demand function, they are likely to estimate them probabilistically. Probabilistic approaches to oligopoly have been initiated by I.Horowitz(1970a), D.Tarr(1972) and R.M.Cyert and M.H.DeGroot (1970b, 1971, 1973). A pioneering intertemporal duopoly model by

J.W.Friedman(1968) is the basis for all of the Cyert-DeGroot works where Bayesian approaches are adopted. In Chapter 9, we shall briefly refer to some salient features of probabilistic oligopoly models.

All oligopoly models alluded to so far may be considered natural extensions of the original Cournot model. Quite different oligopoly models may be constructed on the basis of game theoretic considerations. The reader should refer to M.Shubik(1959), L.Telser(1972) and the most recent and original work by J.W.Friedman(1975) for game theoretic models.

Until quite recently, oligopoly models have been placed in partial equilibrium frameworks. General equilibrium models where oligopolists are assumed to behave 'a la Cournot' have been formulated by J.J.Gabscewicz and J.Vial(1972) and J.Marschak and R.Selten(1974).

Summarily, we shall be concerned in this book with stability analysis of some oligopoly models which may be considered to have evolved from the original Cournot model and which may best be distinguished from one another by differences in forms of expectations that firms are assumed to have on rivals' behaviors.

CHAPTER 1

EXISTENCE AND STABILITY OF THE COURNOT OLIGOPOLY SOLUTION(OR EQUILIBRIUM)

1.1. No Product Differentiation

1.1.1. Existence Proof

One of the most well-known models of oligopoly with or without
product differentiation is that of A.Cournot(1838), an early French
mathematical economist. According to Cournot, each firm in any oligopolistic industry determines its optimal profit maximizing output in any
period on the naive assumption that all of its rivals' outputs will
remain at the same levels as in the immediately preceding period.

In this subsection we shall be concerned with existence and stability of equilibrium in a Cournot model involving no product differentiation (that is, a homogeneous Cournot model). The Cournot oligopoly
solution (or equilibrium) is defined to be a state in which optimal
response of each firm to others' outputs will not induce change in its
output. Such equilibrium does not necessarily exist as is seen in
Fig. 1, which is essentially due to M.McManus(1964a).

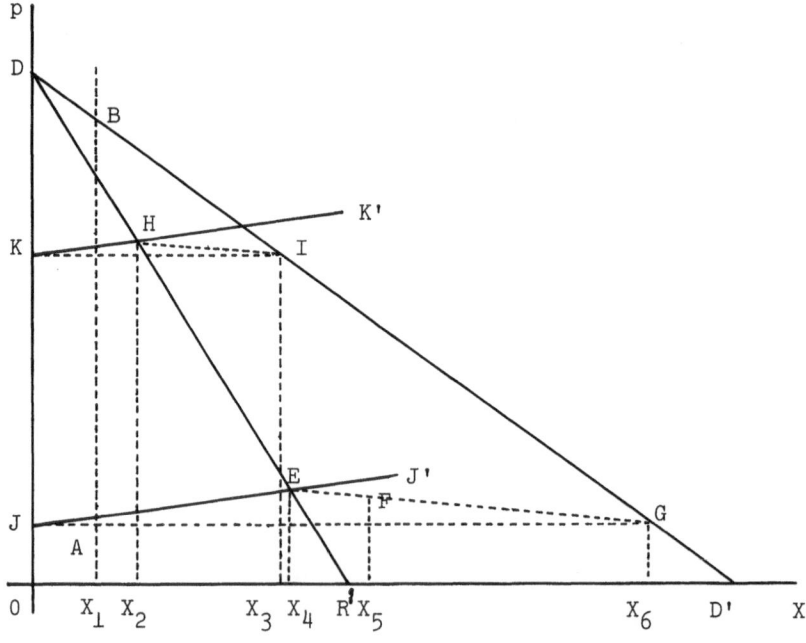

Fig. 1

In Fig. 1 concerned with a duopoly, DD' denotes a market demand curve, DR' a marginal revenue curve of a firm when the other firm's output is zero, JJ' a marginal cost curve of a low cost firm (A for short) and KK' a marginal cost curve of a high cost firm (called B) which exceeds that of A by KJ regardless of output level. Under the Cournot assumption, if B's output in period t-1 is zero, A's profit maximizing output in period t is given by OX_4, and A's marginal revenue and marginal cost are equalized. If B's output in period t-1 is designated OX_1, A's demand curve shifts to BD', giving rise to a corresponding marginal revenue curve starting from B, while the marginal cost curve emanates from A and has the same slope as JJ'. Accordingly, A's profit maximizing output in period t equals X_1X_5. The same marginal revenue and marginal cost corresponding to this output is given by X_5F. The sum of A's output in period t and that of B in period t-1 amounts to OX_5. Similarly, as B's output in period t-1 increases, the points for which A's marginal cost and marginal revenue curves intersect move downward along the line EG, and the sum of the two firms' outputs increases from OX_4 to OX_6. However, in order to make A's output non-negative, B's output can not go beyond OX_6. The same arguments applied to the determination of B's optimal output in period t for a given A output in period t-1 will yield a downward sloping line HI along which B's marginal cost curve and marginal revenue curve intersect, and the sum of the two firms' outputs increases from OX_2 to OX_3. Since B's output can never become negative, A's output must equal or be less than OX_3. For both firms' outputs to be non-negative, X_2X_3 and X_4X_6 must overlap somewhere which is a sheer impossibility. Thus the Cournot oligopoly solution does not exist under the stipulated conditions for Fig. 1. The principal cause for non-existence of the Cournot oligopoly solution is the marked difference between the two firms' (marginal) cost functions.

Existence of non-negative competitive equilibrium in a Walrasian general equilibrium model with production and consumption has been investigated by such economists as K.J.Arrow and G.Debreu(1954), D.Gale(1955), H.Nikaido(1956), L.W.McKenzie(1959). Influenced by works of these economists, rigorous mathematical proofs of existence of the Cournot oligopoly solution have been attempted by E.Burger(1959), M.McManus(1962b, 1964) and C.R.Frank Jr. and R.E.Quandt(1963). Burger gave a non-cooperative game theoretical proof for a simple model in which all firms are assumed to have identical and strictly monotonic increasing cost functions. McManus in his diagrammatic proof was concerned with the case of no production cost for each firm and the same case as Burger. Frank and Quandt's proof based on the Brouwer fixed point theorem, is most general in that it allows for differences in

cost functions among firms. This proof, however, does not necessarily
ensure non-negative profits for firms in equilibrium.

    With these preparatory remarks, we shall now proceed to give an
independent proof of existence of the Cournot oligopoly solution to
ensure that each firm has non-negative profit in equilibrium. Let us
first enumerate notations and assumption. Suffixes run from 1 through
n, n being the number of firms.

__Assumption 1.1.1.1.__: The i-th firm's output x_i belongs to $\Omega_i = [0, M_i]$,
M_i being a positive finite number.

__Assumption 1.1.1.2.__: The market price of homogeneous goods as a
function of the total outputs of all firms, $p = f(\sum_i x_i)$, is
continuous and decreasing in $\sum_i x_i$. Differentiability of f is
not necessary.

__Assumption 1.1.1.3.__: The i-th firm's total cost as a function of its
output, $C_i(x_i)$, is continuous in x_i, and $C_i(0) = 0$.
Differentiability of C_i is not needed.

__Assumption 1.1.1.4.__: The i-th firm's profit function

$$\pi_i(x) \equiv \pi_i(x_i; x_{-i})$$

$$= x_i f(\sum_i x_i) - C_i(x_i)$$

is concave in x_i, where

$$x = (x_1, x_2, \cdots, x_n) \in \Omega \equiv \prod_i \Omega_i,$$

$$x_{-i} = (x_1, \cdots, x_{i-1}, x_{i+1}, \cdots, x_n) \in \Omega_{-i} \equiv \prod_{j \neq i} \Omega_j.$$

__Assumption 1.1.1.5.__:

$$C_i'(0) < f(\sum_{j \neq i} x_j), \quad x \in \Omega$$

$$M_i f'(\sum_{j \neq i} x_j + M_i) + f(\sum_{j \neq i} x_j + M_i) < C_i'(M_i), \quad x \in \Omega.$$

    A few comments on these assumptions might be in order. Since the
differentiability of either f or C_i is unnecessary, kinks as in Figs 2a
and 2b are possible.

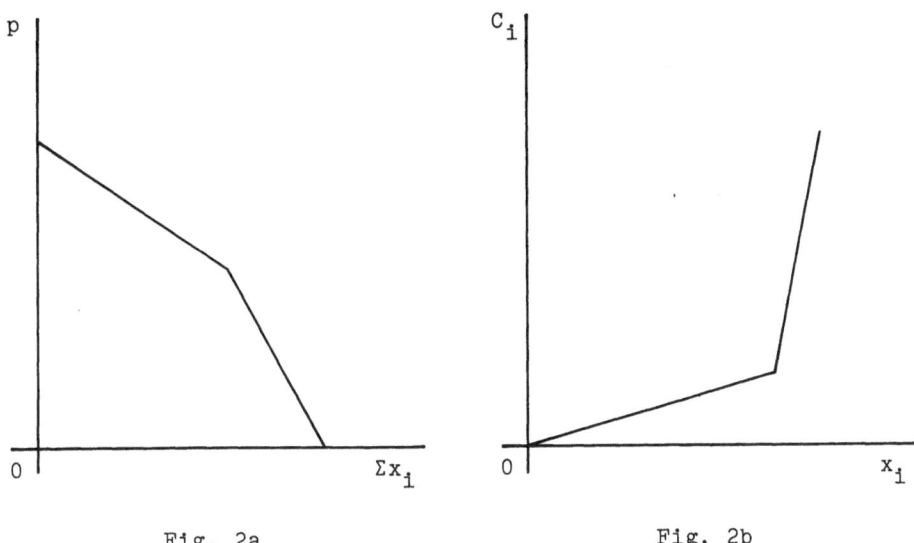

Fig. 2a   Fig. 2b

Assumption 1.1.1.4 which is equivalent to $\partial^2 \pi_i / \partial x_i^2 \leq 0$ ensures the
existence of the i-th firm's profit maximizing output against given
$\sum_{j \neq i} x_j$. It is satisfied provided:

   (a) Marginal revenue of the i-th firm, MR^1, is decreasing in x_i
and the marginal cost of the i-th firm, MC^1, is either constant or
increasing in x_i.

   (b) MR^1 is increasing in x_i and MC^1 increases at the same rate or
faster than MR^1.

   (c) MR^1 is decreasing in x_i, while MC^1 decreases at the same rate
or slower than MR^1.

Assumption 1.1.1.5 is needed to ensure interior maximum.

   We are now able to state and prove

Theorem 1.1.1.1.: Under Assumptions 1.1.1.1 - 1.1.1.5 the (interior)
            Cournot oligopoly solution exists.

Proof: Our proof is divided into two steps.

   1st. We adopt a less general assumption that π_i is strictly concave in
x_i. Let $x_{-i} \in \Omega_{-i}$ be arbitrarily given. Since $\pi_i(x_i; \pi_{-i})$ is continuous
in $x_i \in \Omega_i$, compactness of Ω_i and strict concavity of the profit function
together ensure existence of the i-th firm's profit maximizing output to
be denoted by $h_i(x_{-i})$ as a single valued function of x_{-i}. To prove
continuity of $h_i(x_{-i})$, consider a sequence

$$\{x_{-i}\}_{\tau=1}^{\infty}, \quad \text{where} \quad x_{-i} \in \Omega_{-i},$$

and let

$$\lim_{\tau \to \infty} x_{-i}^{\tau} = x_{-i}^{0}.$$

From compactness of Ω_{-i}, $x_{-i}^{0} \in \Omega_{-i}$. Given the sequence above, we get

$$\{h_i(x_{-i}^{\tau})\}_{\tau=1}^{\infty}, \text{ where } h_i(x_{-i}^{\tau}) \in \Omega_i.$$

As Ω_i is compact, there exists a convergent subsequence of the immediately above sequence. For the sake of notation simplicity, let $\{h_i(x_{-i}^{\tau})\}_{\tau=1}^{\infty}$ be such a subsequence. By definition,

$$(1) \qquad \pi_i(h_i(x_{-i}^{\tau}); x_{-i}^{\tau}) \geq \pi_i(x_i; x_{-i}^{\tau}), \; x_i \in \Omega_i.$$

By virtue of continuity of the profit function,

$$\pi_i(\lim_{\tau \to \infty} h_i(x_{-i}^{\tau}); x_{-i}^{0}) \geq \pi_i(x_i; x_{-i}^{0}), \; x_i \in \Omega_i.$$

Single-valuedness of $h_i(x_{-i})$ yields,

$$\lim_{\tau \to \infty} h_i(x_{-i}^{\tau}) = h_i(x_{-i}^{0}),$$

which shows continuity of $h_i(x_{-i})$.

Consider now a continuous point-to-point mapping from a compact convex set Ω into itself,

$$H(x) = (h_1(x_{-1}), \cdots, h_n(x_{-n})):$$

Application of the Brouwer fixed point theorem[1] enables us to assert existence of at least one x* such that

$$x* = H(x*),$$

where x* is nothing but the Cournot oligopoly solution whose existence was to be established. Incidentally, we may note that each firm earns non-negative profit in equilibrium due to Assumption 1.1.1.3.[2]

2nd. We now weaken strict concavity of profit function to concavity. Let

$$\{\varepsilon_\nu\}_{\nu=1}^{\infty}$$

be a sequence of real numbers such that

$$\varepsilon_\nu > 0, \lim_{\nu \to \infty} \varepsilon_\nu = 0.$$

As π_i is concave in x_i, a function defined by

---

1) See, for example, Nikaido(1968) for this theorem.
2) As was already mentioned, Frank and Quandt's proof does not ensure non-negativity of profit for each firm in equilibrium. Instead of $C_i(0) = 0$, they assume strictly monotone increase of C_i in x_i.

(2) $\pi_i^{\varepsilon_\nu}(x_i; x_{-i}) \equiv \pi_i(x_i; x_{-i}) - \varepsilon_\nu x_i^2$

is strictly concave in x_i. Applying the result attained in the 1st step to the modified profit function (2), we can assert for any ε_ν existence of at least one fixed point $x^*(\varepsilon_\nu) \in \Omega$. Let a sequence of fixed points be given by

$$\{x^*(\varepsilon_\nu)\}_{\nu=1}^{\infty}$$

By definition of the fixed point,

(3) $\quad x^*(\varepsilon_\nu) = H^{\varepsilon_\nu}(x^*(\varepsilon_\nu))$

$\qquad\qquad = (h_1^{\varepsilon_\nu}(x_{-1}^*(\varepsilon_\nu)), \cdots, h_n^{\varepsilon_\nu}(x_{-n}^*(\varepsilon_\nu)))$.

As Ω is compact, the sequence of the fixed point has a convergent subsequence. Let the original sequence denote such subsequence to avoid complicated notation. From (2),

(4) $\quad \lim_{\nu \to \infty} \pi_i^{\varepsilon_\nu}(x_i^*(\varepsilon_\nu); x_{-i}^*(\varepsilon_\nu)) = \pi_i(\lim_{\nu \to \infty} h_i^{\varepsilon_\nu}(x_{-i}^*(\varepsilon_\nu)); \lim_{\nu \to \infty} x_{-i}^*(\varepsilon_\nu))$,

(5) $\quad \lim_{\nu \to \infty} \pi_i^{\varepsilon_\nu}(x_i; x_{-i}^*(\varepsilon_\nu)) = \pi_i(x_i; \lim_{\nu \to \infty} x_{-i}^*(\varepsilon_\nu)), \ x_i \in \Omega_i$.

Thus we have

$$\pi_i(\lim_{\nu \to \infty} h_i^{\varepsilon_\nu}(x_{-i}^*(\varepsilon_\nu)); \lim_{\nu \to \infty} x_{-i}^*(\varepsilon_\nu)) \geq \pi_i(x_i; \lim_{\nu \to \infty} x_{-i}^*(\varepsilon_\nu)), x_i \in \Omega_i,$$

which shows that

$$\lim_{\nu \to \infty} x^*(\varepsilon_\nu) = \lim_{\nu \to \infty} H^{\varepsilon_\nu}(x^*(\varepsilon_\nu))$$

is the Cournot oligopoly solution. Q.E.D.

1.1.2. Stability Analysis

R.Theocharis(1960) in his seminal paper provoking lively discussions on stability of the Cournot oligopoly solution concluded that under the assumption of constancy of the marginal cost of each firm, the downward sloping demand curve coupled with instantaneous adjustment of each firm's actual output to a profit maximizing one will ensure stability of the Cournot oligopoly solution in a discrete (difference equation) system only in the case of duopoly. That Theocharis' conclusion is not generally tenable was pointed out by F.M.Fisher(1961), M.McManus and R.E.Quandt (1961) and R.L.Bishop(1962). F.H.Hahn(1962) gave an excellent proof of stability in a continuous (differential equation) system by use of Lyapunov's second method. His result was greatly extended by Okuguchi (1964).

In this subsection we shall first analyze the stability of the

Cournot oligopoly solution in a discrete dynamic system. Let $x_i^*(t)$ be the i-th firm's profit maximizing output in period t against the rest of industry output in period t-1, $x^1(t-1) \equiv \sum_{j \neq i} x_j(t-1)$. The first order and second order conditions for the i-th firm's profit maximization under the Cournot behavioristic assumption are:

(1) $f(x^1(t-1) + x_i^*(t)) + x_i^*(t)f'(x^1(t-1) + x_i^*(t))$

$- C_i'(x_i^*(t)) = 0,$ $i=1,2,\cdots,n,$

(2) $2f'(x^1(t-1) + x_i^*(t)) + x_i^*(t)f''(x^1(t-1) + x_i^*(t))$

$-C_i''(x_i^*(t)) < 0,$ $i=1,2,\cdots,n,$

where $x^1 \equiv \sum_{j \neq i} x_j$.

From (1) we get

(3) $x_i^*(t) = g_i(x_1(t-1),\cdots,x_{i-1}(t-1),x_{i+1}(t-1),\cdots,x_n(t-1))$

$= h_i(x^1(t-1)),$ $i=1,2,\cdots,n,$

(4) $\partial x_i^*(t)/\partial x_{j-1}(t-1) = (f' + x_i^*(t)f'')/(2f' + x_i^*(t)f'' - C_i''),$

$i \neq j, \; i,j = 1,2,\cdots,n,$

where it is assumed that g_i is single valued and differentiable. As is easily seen, the numerator in (4) is the rate of change of the i-th firm's marginal revenue MR^1 with respect to change in x_j or the rest of industry output. We assume that it is negative. Actual output changes according to [1]

(5) $x_i(t) - x_i(t-1) = k_i(x_i^*(t) - x_i(t-1)),$ $0 < k_i \leq 1,$

$i=1,2,\cdots,n.$

From (3) and (5), and taking into consideration the mean value theorem for a function of several variables we derive,

(6) $|x_i(t+1) - x_i(t)| \leq \sum_{j \neq i} k_i|\alpha_{ij}||x_j(t) - x_j(t-1)|$

$+ (1-k)|x_i(t) - x_i(t-1)|,$ $i=1,\cdots,n,$

where $\alpha_{ij} = \partial g_i/\partial x_j(t-1)$ $i \neq j, \; i,j = 1,2,\cdots,n$

---

[1] In the case of Theocharis' instantaneous adjustment $k_i=1$ for all i. Fisher(1961), on the other hand, assumes only the positivity of k_i for all i. Thus inequality (7) is replaced by
$$\sum_{j \neq i} k_i|\alpha_{ij}| + |1-k_i| < 1, \quad i=1,2,\cdots,n.$$

is evaluated at a value between $x(t-1)$ and $x(t)$, thus depending on time. To proceed further, assume that

(7) $\qquad \sum_{i \neq j} k_i |\alpha_{ij}| - k_j < 0, \qquad j=1,2,\cdots,n,$

and define α which is positive and less than one by

$$\alpha \equiv \max_j (\sum_{i \neq j} k_i |\alpha_{ij}| - k_j) + 1.$$

From (6) and (7),

$$\sum_i |x_i(t+1) - x_i(t)| \leq \alpha \sum_i |x_i(t) - x_i(t-1)|,$$

which shows that a mapping defined as

$$(x_1(t),\cdots,x_n(t)) = (k_1 \, h_1(t-1)) + (1-k)x_1(t-1),\cdots,$$

$$k_n \, h_n(x^n(t-1)) + (1-k)x_n(t-1)): \quad \Omega \to \Omega$$

is a contraction with the contraction constant $\alpha^{1)}$. Thus it has a unique fixed point -the Cournot oligopoly solution- which is approached from any initial condition, establishing global stability.

The fundamental inequality (7) which is our global stability condition may be interpreted as follows. Differentiating the i-th firm's marginal revenue with respect to x_i^* and x_j, respectively,

$$MR_i^i \equiv \partial MR^i / \partial x_i^* = 2f' + x_i^* f'' - C_i'' < 0, \qquad i=1,2,\cdots,n,$$

$$MR_j^i \equiv \partial MR^i / \partial x_j = f' + x_i^* f'' < 0, \qquad i,j=1,2,\cdots,n, \; i \neq j,$$

in the light of which (7) is rewritten as

(8) $\qquad \sum_{i \neq j} k_i MR_j^i / (MR_i^i - C_i'') < k_j, \qquad j=1,2,\cdots,n.$

From this, one sees that _ceteris paribus_, the greater the absolute value of MR_i^i, the more likely (8) will be satisfied, provided $C_i'' \geq 0$. Consider a simple case where

$$k_i = 1, \; i = 1,2,\cdots,n,$$

$$p = a - b \sum_i x_i, \; a, \, b > 0$$

$$C_i = e_i + c_i x_i, \; i = 1,2,\cdots,n.$$

---

1) See Mathematical Appendix for a theorem on contraction. In our case, the metric between two vectors x and y is defined by

$$\sigma(x, y) \equiv \sum_i |x_i - y_i|.$$

We may easily see that (8) is satisfied if n=2, confirming Theocharis'
conclusion referred to above. Where the i-th firm has a quadratic cost
function

$$C_i = e_i + c_i x_i + d x_i^2/2, \qquad i=1,2,\cdots,n$$

and demand function is linear as in the above case, the stability
condition (8) reduces to

$$b \sum_{i \neq j} k_i/(2b + d) < k_j, \qquad j=1,2,\cdots,n.$$

From this one can conclude:

(a) <u>Ceteris paribus</u>, increase in n is more likely to dissatisfy
the stability condition.

(b) <u>Ceteris paribus</u>, increase in d is more likely to satisfy the
stability condition.

(c) <u>Ceteris paribus</u>, increase in b is more likely to satisfy the
stability condition.

We next turn to stability analysis of continuous systems. We shall
assume, for avoidance of complication, uniqueness of the Cournot oligopoly
solution.[1] Our proofs are based on the following fundamental stability
theorem for a continuous dynamic system.[2]

<u>Theorem 1.1.2.1</u>: Consider a dynamic system of differential equations,

$$(*) \quad dy/dt = f_i(y), \qquad i=1,2,\cdots,n,$$

and assume that f_i's are continuous in y. Let T be a compact
set in n-dimensional Euclidean space R^n and

$$y^0 = (y_1^0, y_2^0, \cdots, y_n^0) \in T$$

be a given initial value. Assume further that:

(a) (*) has a unique and continuous solution $y(t; y^0)$ with
respect to y^0.

(b) $y(t; y^0) \in T$ for all t.

(c) There exists a function of time

$$V(t) = V(y(t; y^0))$$

which is continuous in y and strictly monotonic decreasing in
t unless $y \neq y^*$, where

---

1) For an explicit treatment of uniqueness of the Cournot oligopoly
solution, the reader should refer to Chapter 2.
2) See Uzawa(1961). The function V is called a (modified) Lyapunov
function.

$$f_i(y^*) = 0, \qquad i=1,2,\cdots,n$$

and y* is unique.

Under these assemptions y* is globally stable, that is, y converges to y* irrespective of the value of the initial condition.

We have already proved that under Assumptions 1.1.1.1 - 1.1.1.5 the interior Cournot oligopoly solution exists. The solution, however, may not be unique. To proceed to analysis of stability for continuous systems, define

$$\bar{x}_i \equiv x_i^* - x_i, \quad x^i \equiv \sum_{j \neq i} x_j$$

to derive the first order and second order conditions for profit maximization,

(9) $\qquad f(x_i^* + x^i) + x_i^* f'(x_i^* + x^i) - C_i'(x_i^*) = 0, \qquad i=1,2,\cdots,n,$

(10) $\qquad 2f'(x_i^* + x^i) + x_i^* f''(x_i^* + x^i) - C_i''(x_i^*) < 0, \qquad i=1,2,\cdots,n.$

As for adjustment of output we assume

(11) $\qquad dx_i/dt = F_i(\bar{x}_i), \text{ sgn } F_i = \text{sgn } \bar{x}_i, \qquad i=1,2,\cdots,n,$

where F_i is a sign-preserving, single-valued and continuous function of $\bar{x}_i$ with $F_i(0) = 0$ for all i. The sign-preserving property means that the rate of change of actual output of each firm has the same sign as the difference between its profit maximizing output and the actual one. This hypothesis is, in spirit, in accord with the adjustment hypothesis adopted in a recent analysis of stability of the competitive analysis in a Walrasian general equilibrium model, where the rate of price change of any goods in tâtonnement process is assumed to have the same sign as its excess demand. It should be noted, however, that Theocharis, Fisher, Bishop, McManus and Quandt and Hahn all adopted proportional adjustment hypothesis, that is,

(12) $\qquad dx_i/dt = k_i \bar{x}_i, \qquad k_i > 0, i=1,2,\cdots,n.$

Differentiate (9) with respect to time to derive,

(13) $\qquad dx_i^*/dt = -\alpha_i dx^i/dt, \qquad i=1,2,\cdots,n,$

where

(14) $\qquad \alpha_i \equiv (f' + x_i^* f'')/(2f' + x_i^* f'' - C_i''), \qquad i=1,2,\cdots,n.$

Define,

$$D_i(\bar{x}_i) \equiv d(\bar{x}_i F_i(\bar{x}_i))/d\bar{x}_i$$

$$= F_i(\bar{x}_i) + \bar{x}_i F_i'(\bar{x}_i), \qquad i=1,2,\cdots,n.$$

Uniqueness of the Cournot oligololy solution is assumed in all of the following theorems. To analyze stability the following assumptions have to be introduced.

Assumption 1.1.2.1: $f' + x_i f'' < 0$, $x \in \Omega$, $\qquad i=1.2.\cdots,n$

Assumption 1.1.2.2: $f' < C_i''$, $x \in \Omega$, $\qquad i=1,2,\cdots,n.$

Assumption 1.1.2.3: $f'' = 0$, $C_i'' = d > 0$, $\qquad i=1,2,\cdots,n.$

Assumption 1.1.2.4: $F_i' > 0$, $\qquad i=1,2,\cdots,n.$

Assumption 1.1.2.5: $\sum_i F_i \sum_i D_i \geq 0.$

With these preparations we can state and prove

Theorem 1.1.2.2: Under Assumptions 1.1.2.1. - 1.1.2.5, the Cournot oligopoly solution is globally stable.

Proof: Note first that Assumptions 1.1.2.1. - 1.1.2.2. together with (10) imply $0 < \alpha_i < 1$, and let $\alpha_i = \alpha$ for all i to take into account Assumption 1.1.2.3. Define a Lyapunov function V by

$$V \equiv \sum_i \bar{x}_i F_i(\bar{x}_i),$$

which is positive unless $\bar{x}_i \neq 0$ for all i. Differentiating V with respect to time,

$$dV/dt = \sum_i d\bar{x}_i/dt F_i + \sum_i \bar{x}_i F_i' d\bar{x}_i/dt$$

$$= \sum_i dx_i^*/dt F_i - \sum_i F_i^2 + \sum_i \bar{x}_i F_i' dx_i^*/dt - \sum_i \bar{x}_i F_i F_i'$$

$$= -\{\sum_i F_i D_i + \alpha \sum_i D_i dx^i/dt\}.$$

The first term in the brace is non-negative. Suppose that the second term is negative, and take into consideration $0 < \alpha < 1$ and Assumption 1.1.2.5 to conclude that the expression in the brace is greater than or equal to $\sum_i F_i \sum_i D_i$ which is non-negative, hence $dV/dt < 0$ unless $\bar{x}_i \neq 0$ for all i. Invoking Theorem 1.1.2.1 global stability is proved. Q.E.D.

A few comments on this theorem are in order. In the case of proportional adjustment hypothesis as given by (12), Assumptions 1.1.2.4 - 1.1.2.5 are satisfied. In the case where $\bar{x}_i \geq 0$ for all i always or $\bar{x}_i \leq 0$ for all i always, Assumption 1.1.2.5 holds true also. Note that stability here is independent of the values of k_i's, which contrasts sharply with a discrete system as seen in stability condition (7) or (8).

To analyze a case where α_i' are not necessarily the same, define a subset I of N = {1,2,$\cdots$,n} and its complimentary set I^c as

$$I \equiv \{i | F_i dx^i / dt < 0\}.$$

$$I^c \equiv N - I.$$

The subset I may be empty, and both I and I^c depend on time. Let us introduce,

Assumption 1.1.2.6: $\quad \sum_{i \in I} D_i \sum_{j \in I^c} F_j \geq 0$

Assumption 1.1.2.7: $\quad \sum_{i \in I} \bar{x}_i F_i' \sum_{i \in I} F_i \geq 0.$

Theorem 1.1.2.3: Under Assumptions 1.1.2.1 - 1.1.2.2, 1.1.2.4, 1.1.2.6 - 1.1.2.7, the Cournot oligopoly solution is globally stable.

Proof: Define V by

$$V \equiv \sum_i x_i F_i,$$

and take into account $0 < \alpha_i < 1$ for all i to get,

$$dV/dt = -\{\sum_i F_i^2 + \sum_i \alpha_i F_i dx^i/dt + \sum_i \alpha_i \bar{x}_i F_i' dx^i/dt + \sum_i \bar{x}_i F_i F_i'\}$$

$$\leq -\{\sum_{i \in I} F_i^2 + \sum_{i \in I} F_i dx^i/dt + \sum_{i \in I} \bar{x}_i F_i' dx^i/dt + \sum_{i \in I} \bar{x}_i F_i F_i' + A\}$$

$$= -\{\sum_{i \in I} F_i^2 + \sum_{i \in I}\sum_{j \neq i} F_i F_j + \sum_{i \in I} \bar{x}_i F_i F_i' + \sum_{i \in I}\sum_{j \neq i} \bar{x}_i F_i' F_j + A\}$$

$$= -\{\sum_{i \in I} F_i \sum_j F_j + \sum_{i \in I} \bar{x}_i F_i' \sum_j F_j + A\}$$

$$= -\{(\sum_{i \in I} F_i)^2 + \sum_{i \in I} F_i \sum_{j \in I^c} F_j + \sum_{i \in I} \bar{x}_i F_i' \sum_{j \in I^c} F_j$$

$$+ \sum_{i \in I} \bar{x}_i F_i' + \sum_{i \in I} F_i + A\}$$

$$= -\{(\sum_{i \in I} F_i)^2 + \sum_{i \in I} D_i \sum_{j \in I^c} F_j + \sum_{i \in I} \bar{x}_i F_i' \sum_{i \in I} F_i + A\},$$

where $\quad A \equiv \sum_{j \in I^c} F_j^2 + \sum_{j \in I^c} \bar{x}_j F_j F_j' + \sum_{j \in I^c} \alpha_j F_j dx^i/dt + \sum_{j \in I^c} \alpha_j \bar{x}_j F_j' dx^j/dt \geq 0.$

Assumption 1.1.2.6 - 1.1.2.7 thus lead to dV/dt < 0 provided $\bar{x}_i \neq 0$ for at least one i. Hence, global stability. $\qquad$ Q.E.D.

If $\bar{x}_i \geq 0$ for all i or $\bar{x}_i \leq 0$ for all i during adjustment process, I becomes empty and global stability follows. Assumption 1.1.2.7 is fulfilled for a proportional adjustment hypothesis (12). Assumption

1.1.2.6, however, is purely a mathematical one to which economic interpretation is hard to give.

In Theorems 1.1.2.2 - 1.1.2.3 strictly monotonic increasingness
of F_i is commonly assumed. In the next two theorems this assumption
is dropped. We need, however,

<u>Assumption 1.1.2.8</u>: $\sum_i \bar{x}_i \sum_i F_i \geq 0$.

<u>Theorem 1.1.2.4</u>: Under Assumptions 1.1.2.1 - 1.1.2.3 and 1.1.2.8 the
Cournot oligopoly solution is globally stable.

<u>Proof</u>: Let

$$V \equiv \sum_i \bar{x}_i^2/2$$

be a Lyapunov function. Differentiating it with respect to time,

$$dV/dt = \sum_i \bar{x}_i(dx_i^*/dt - dx_i/dt)$$

$$= -\{\sum_i \bar{x}_i F_i + \alpha \sum_i \sum_{j \neq i} \bar{x}_i F_j\},$$

where $\alpha_i = \alpha < 1$ for all i. As the first term in the brace is non-
negative, the whole expression in the brace becomes non-negative if the
second term is so. Where the second term is negative, we have

$$dV/dt \leq -\{\sum_i \bar{x}_i F_i + \sum_i \sum_{j \neq i} \bar{x}_i F_j\}$$

$$= - \sum_i \bar{x}_i \sum_i F_i \leq 0,$$

leading to global stability. Q.E.D.

If $\bar{x}_i \geq 0$ or $\bar{x}_i \leq 0$ for all i during the whole adjustment process
(inequality sign in the same direction), Assumption 1.1.2.8 is surely
met. In the case of proportional adjustment described by (12), the
same assumption holds true if k_i's are the same.

In the next theorem the restrictive Assumption 1.1.2.3 is dropped
and Assumption 1.1.2.8 is replaced by

<u>Assumption 1.1.2.9</u>: $\sum_{i \in I'} \bar{x}_i \sum_i F_i \geq 0$,

where

$$I' = \{i \mid \bar{x}_i \sum_{j \neq i} F_j < 0\}$$

$$I'^C = N - I'.$$

<u>Theorem 1.1.2.5</u>: Under Assumptions 1.1.2.1, 1.1.2.2 and 1.1.2.9 the
Cournot oligopoly solution is globally stable.

<u>Proof</u>: Let V be defined by

$$V = x_i^2/2,$$

from which

$$dV/dt = -\{\sum_i \bar{x}_i F_i + \sum_i \sum_{j \neq i} \alpha_i \bar{x}_i F_j\}$$

$$\leq -\{\sum_i \bar{x}_i F_i + \sum_{i \in I'} \bar{x}_i \sum_{j \neq i} F_j + \sum_{j \in I'^c} \sum_{m \neq j} \alpha_j \bar{x}_j F_m\}$$

$$= -\{\sum_{i \in I'} \bar{x}_i \sum_i F_i + \sum_{i \in I'^c} \bar{x}_i F_i + \sum_{j \in I'^c} \sum_{m \neq j} \alpha_j \bar{x}_j F_m\} \leq 0.$$

Q.E.D.

Theorem 1.1.2.5 holds true if $\bar{x}_i \geq 0$ for all i or $\bar{x}_i \leq 0$ for all i in the process of adjustment. It is, however, rather difficult to give economic interpretation to Assumption 1.1.2.9.

1.2. Product Differentiation

In this section we shall be concerned with existence and stability of the Cournot oligopoly solution in a model with product differentiation. Thus n firms exist, producing close substitutes for one another. The following assumption are necessary for the proof of existence.

Assumption 1.2.1: The i-th firm's output x_i belongs to a compact interval S_i containing zero.

Assumption 1.2.2: p_i, price of the i-th goods which is produced by the i-th firm, is a continuous function of x_i's,

$$p_i = f^1(x_1, x_2, \cdots, x_n).$$

Assumption 1.2.3: The i-th firm's cost function is continuous in x_i and $C_i(0) = 0$.

Assumption 1.2.4: The i-th firm's profit function

$$\pi_i(x) = x_i f^1(x_1, x_2, \cdots, x_n) - C_i(x_i)$$

is concave in x_i.

Our existence theorem is stated as

Theorem 1.2.1: Under Assumptions 1.2.1 - 1.2.4 the Cournot oligopoly solution in a model with product differentiation exists.

Proof: To prove this by use of the Kakutani fixed point theorem, let us define

$$x = (x_1, x_2, \cdots, x_n) \in S \equiv \prod_i S_i$$

$$x_{-i} = (x_1, \cdots, x_{i-1}, x_{i+1}, \cdots, x_n) \in \prod_{j \neq i} S_j.$$

As π_i is concave in x_i and S_i compact, there exists x_i which maximizes $\pi_i(x_i, x_{-i})$ against given x_{-i}. Denote maximizing element(s) by $h_i(x_{-i})$, and introduce a mapping

$$H(x) = (h_1(x_{-1}), \cdots, h_n(x_{-n})): \quad S \to S.$$

Due to Assumption 1.2.4, the image of $H(x)$ is convex. We now claim
upper semi-continuity of $H(x)$. Consider two sequences

$$\{x^\tau\}_{\tau=1}^\infty, \quad \{y^\tau\}_{\tau=1}^\infty$$

such that

$$y^\tau \in H(x^\tau), \qquad \tau = 1, 2, \cdots, \infty$$

and let

$$\lim_{\tau\to\infty} x^\tau = x^0 \in S, \quad \lim_{\tau\to\infty} y^\tau = y^0 \in S.$$

If $H(x)$ is not upper semi-continuous,

$$\pi_1(y_i^0, x_{-i}^0) < \pi_1(h_i(x_{-i}^0), x_{-i}^0)$$

has to hold for at least one i. Continuity of π_i implies

$$\pi_1(y_i^\tau, x_{-i}^\tau) < \pi_1(h_i(x_{-i}^0), x_{-i}^\tau)$$

for at least one i and for a sufficiently large τ. For such i and τ

$$y_i^\tau \notin h_i(x_{-i}^\tau),$$

contradicting to our supposition $y^\tau \in H(x^\tau)$ for all τ. As $H(x)$ satisfies
all conditions for the Kakutani fixed point theorem[1], there must exist at
least one x* such that x* $\in H(x*)$, completing proof. Q.E.D.

As for stability, Hadar(1966), on one hand, proved stability of the
Cournot oligopoly solution in a discrete model involving product differentiation only for instantaneous adjustment of actual output of each
firm to a profit maximizing one. Quandt(1967), on the other hand, was
concerned with stability in a continuous model of price adjustment.

We shall first deal with stability of the Cournot oligopoly solution in a discrete quantity adjustment model not necessarily allowing
for instantaneous adjustment. Under the Cournot assumption, the i-th
firm's profit maximizing output in period t, $x_i^*(t)$, is determined by
the following equation.

(1) $\quad f^i(x_i^*(t), x_{-i}(t-1)) + x_i^*(t) f_i^i(x_i^*(t), x_{-i}(t-1))$

$$- C_i'(x_i^*(t)), \qquad i = 1, 2, \cdots, n,$$

where interior maximum is assumed and

---

1) For the Kakutani fixed point theorem, see Kakutani(1941), Nikaido
(1968) and Morishima(1969).

$$f_i^i \equiv \partial f^i / \partial x_i^*(t), \qquad i=1,2,\cdots,n.$$

From (1) we derive

$$(2) \qquad \partial x_i^*(t)/\partial x_j(t-1) = -(f_i^i + x_i^*(t)f_{ij}^i)/(2f_i^i + x_i^*(t)f_{ii}^i - C_i''),$$

$$i \neq j, \ i, \ j = 1,2,\cdots,n,$$

where

$$f_{ii}^i \equiv \partial f_i^i/\partial x_i^*(t)$$

$$f_{ij}^i \equiv \partial^2 f^i/\partial x_j(t-1)\partial x_i^*(t)$$

$$i \neq j, \ i, \ j = 1,2,\cdots,n.$$

Actual output is assumed to change according to

$$(3) \qquad x_i(t) - x_i(t-1) = k_i(x_i^*(t) - x_i(t-1)), \quad 0 \leq k_i \leq 1, \ i=1,\cdots,n.$$

Solving (1) for $x_i^*(t)$ and assuming that it is single valued and differentiable we have,

$$(4) \qquad x_i^*(t) = g_i(x_{-i}(t-1)) \qquad i=1,2,\cdots,n.$$

Arguing similarly as in derivation of stability condition (7) in Subsection 1.1.2, the Cournot oligopoly solution under product differentiation is unique and globally stable provided,[1])

$$(5) \qquad \underset{i \neq j}{\Sigma} \, k_i |\partial x_i^*(t)/\partial x_j(t-1)| - k_j < 0, \qquad j=1,2,\cdots,n.$$

To elaborate on economic interpretation of stability condition (5), let MR^i denote marginal revenue of the i-th firm with respect to change in its own output,

$$MR^i \equiv f^i + x_i^*(t)f_i^i, \qquad i=1,2,\cdots,n,$$

and define MR_j^i and MR_i^i by

$$MR_j^i \equiv \partial MR^i/\partial x_j(t-1) = f_j^i + x_i^*(t)f_{ij}^i, \qquad i \neq j, \ i, \ j = 1,\cdots,n,$$

$$MR_i^i \equiv \partial MR^i/\partial x_i^*(t) = 2f_i^i + x_i^*(t)f_{ii}^i, \qquad i=1,2,\cdots,n.$$

With an additional
<u>Assumption 1.2.5</u>: $MR_j^i < 0$, $i \neq j$, i, $j = 1,2,\cdots,n$,
the stability condition may be rewritten as

$$(6) \qquad \underset{i \neq j}{\Sigma} \, k_i MR_j^i/(MR_i^i - C_i'') < k_j, \qquad j=1,2,\cdots,n.$$

---

1) If only $k_i > 0$ is assumed for all i, (5) is to be replaced by

$$\underset{i \neq j}{\Sigma} \, k_i |\partial x_i^*(t)/\partial x_j(t-1)| + |1-k_j| < 1, \text{ for all } j.$$

which is formally the same as (8) in 1.1.2. The reader is urged to examine (6) in detail for a simple case of linear demand and quadratic cost functions,

$$p_i = a_{10} + \sum_j a_{ij} x_j, \quad a_{ij} < 0, \; i,j = 1,2,\cdots,n,$$

$$C_i = e_i + c_i x_i + d_i x_i^2/2, \quad i=1,2,\cdots,n.$$

Turning now to a continuous model under product differentiation, the first order condition for profit maximization is given by

(7) $\qquad f^1(x_i^*, \; x_{-i}) + x_i^* f_i^1(x_i^*, \; x_{-i}) - C_i'(x_i^*) = 0, \qquad i=1,\cdots,n.$

Differentiating this with respect to time and rearranging,

(8) $\qquad dx_i^*/dt = - \sum_{j \neq i} \alpha_{ij} dx_j/dt, \qquad i=1,2,\cdots,n,$

where

$$\alpha_{ij} \equiv (f_j^1 + x_i^* f_{ij}^1)/(2f_i^1 + x_i^* f_{ii}^1 - C_i''), \quad i \neq j, \; i,j = 1,\cdots,n$$

and α_{ij}'s are positive by virtue of Assumption 1.2.4 and 1.2.5.[1] As in a model of no product differentiation, actual output is adjusted by

(9) $\qquad dx_i/dt = F_i(\bar{x}_i), \quad \text{sgn } \bar{x}_i = \text{sgn } F_i, \qquad i=1,\cdots,n$

where $\bar{x}_i \equiv x_i^* - x_i, \qquad i=1,2,\cdots,n.$

Let two subsets of $N = \{1,2,\cdots,n\}$ be defined by

$$I = \{i \,|\, D_i \leq 0\}, \quad J = \{j \,|\, F_j \leq 0\}.$$

where $D_i \equiv F_i + \bar{x}_i F_i', \qquad i=1.2.\cdots,n.$ I and J may be empty.
By definition,

$$I^c = N - I, \quad J^c = N - J.$$

When F_i's are monotonic increasing, the following theorem is established.

Theorem 1.2.2: Assume uniqueness of the Cournot oligopoly solution,
$\qquad F_i' > 0$ for all i and $\alpha_{ij} < 1$, $\; i \neq j, \; i,j = 1,2,\cdots,n.$ Then
the solution has global stability property provided:

$$(10) \qquad \sum_{i \in I} D_i (F_i + \sum_{\substack{j \in J^c \\ j \neq i}} F_j) \geq 0,$$

---

1) Note that $x_i^*(t)$ and $x_j(t-1)$ in a discrete system corresponds, respectively, to x_i^* and x_j in a continuous model.

$$(11) \quad \sum_{i \in I^c} D_i(F_i + \sum_{\substack{j \in J \\ j \neq i}} F_j) \geq 0.$$

Proof: Omitted. (Define a Lyapunov function by $V = \sum_i \bar{x}_i F_i$.)

The next theorem is derived without assuming monotonicity of F_i's. Define,

$$I' = \{i \mid x_i \leq 0\}, \quad I'^c = N - I,$$

$$J' = \{j \mid F_j \leq 0\}, \quad J'^c = N - I.$$

Since sgn F_i = sgn x_i, two subsets I' and J' are the same.

Theorem 1.2.3: Assume uniqueness of the Cournot oligopoly solution and $\alpha_{ij} < 1$, $i \neq j$, $i,j = 1,2,\cdots,n$. Then the solution is globally stable provided:

$$(12) \quad \sum_{i \in I'} \bar{x}_i(F_i + \sum_{\substack{j \in J'^c \\ j \neq i}} F_j) \geq 0$$

$$(13) \quad \sum_{i \in I'^c} \bar{x}_i(F_i + \sum_{\substack{j \in J' \\ j \neq i}} F_j) \geq 0.$$

Proof: We omit the proof. (Define $V = \sum_i \bar{x}_i^2/2$.)

A few comments on the above theorems follow. As the numerator and denominator of α_{ij} are MR_j^i and $MR_i^i - C_i''$, respectively, $\alpha_{ij} < 1$ if either

$$(14) \quad C_i'' \geq 0, \quad MR_i^i < MR_j^i, \quad i \neq j, \quad i=1,2,\cdots,n,$$

or

$$(15) \quad f_i^i < C_i'', \quad f_i^i + x_i^* f_{11}^i \leq f_j^i + x_i^* f_{1j}^i, \quad i \neq j, \quad i,j = 1,2,\cdots,n.$$

Inequalities are satisfied whenever $\bar{x}_i \geq 0$ for all i or $\bar{x}_i \leq 0$ for all i in the process of attainment of equilibrium.

1.3. Mathematical Appendix

1.3.1. A Theorem on Contraction Mapping

So far our proofs of stability of the Cournot oligopoly solution in discrete systems have been based on a theorem on contraction mapping for which the reader should refer to, e.g. Kolmogoroff and Fomin(1957). One of the earliest examples of application of the contraction mapping principle in economics is seen in Gale(1964). As for other applications in context of oligopoly models, we may cite Hadar(1965, 1966, 1968), Okuguchi(1969, 1971) and J.W.Friedman(1972, 1975).

Let S be a metric space with metric between two points, x, y $\in$ S, denoted by $\rho(x, y)$. If for a continuous mapping $f(x)$: $S \to S$, an inequality

$$\rho(f(x), f(y)) \le \alpha \rho(x, y)$$

holds for a positive constant $\alpha < 1$, the mapping is said to be a contraction with the contraction constant α. The following theorem is well, known in mathematical literature.

Theorem 1.3.1.1: Let S be a complete metric space which is closed and in which every Cauchy sequence has a limit. If $f(x)$: $S \to S$ is a continuous contraction mapping, then a unique fixed point x* which satisfies

$$x^* = f(x^*), \qquad x^* \in S$$

exists.

Proof: Let x_0 be an arbitrary given initial value, and

$$x_{t+1} = f(x_t), \qquad t=1,2,\cdots,\infty.$$

By assumption,

$$\rho(x_{t+1}, x_t) \le \alpha \rho(x_t, x_{t-1}) \le \cdots \le \alpha^t \rho(x_1, x_0)$$

$$t=1,2,\cdots,\infty.$$

For s such that s > t,

$$\rho(x_s, x_t) \le \rho(x_s, x_{s-1}) + \rho(x_{s-1}, x_{s-2}) + \cdots + \rho(x_{t+1}, x_t)$$

$$\le (\alpha^{s-1} + \alpha^{s-2} + \cdots + \alpha^t)\rho(x_1, x_0)$$

$$= \alpha^t(1 - \alpha^{s-t})\rho(x_1, x_0)/(1 - \alpha)$$

$$\le \alpha^t \rho(x_1, x_0)/(1 - \alpha).$$

Thus a sequence $\{x_\tau\}_{\tau=0}^{\infty}$ is a Cauchy sequence, and there exists x* such that $\lim_{\tau \to \infty} x_\tau = x^* \in S$. By continuity of f, $x^* = f(x^*)$, showing that x* is a fixed point. To show uniqueness, let x* and x** be distinct fixed points:

$$x^* = f(x^*), \qquad x^{**} = f(x^{**}).$$

Thus we get,

$$\rho(x^*, x^{**}) \le \alpha \rho(x^*, x^{**})$$

leading to a contradiction as $0 < \alpha < 1$.

1.3.2. Global Stability of a Discrete System

Though theorems developed in this subsection are not needed directly, they are extremely powerful and convenient in studying global stability of economic systems described by difference equations.

Let a non-autonomous system of difference equations be given by

(1) $x_{t+1} = f(x_t)$, f: $R^n \to R^n$, $f(0) = 0$, $f \in c^1$,

where $f \in c^p$ means that f is p-times continuously differentiable. Brock
and Scheinkman(1975) have derived the following theorem on global stability of the origin of (1).

<u>Theorem 1.3.2.1</u>: If $\| J_f(0)x \| < \| x \|$ for any $x \neq 0$ and if
          $0 \neq \|x\| = f(x)$ implies $\| J_f(x)x \| < \| x \|$, then the origin
          is globally stable where, $\| x \| = (\Sigma x_i^2)^{1/2}$ stands for the
          Euclidean norm and $J_f(x)$ is the Jacobian matrix evaluated at
          x.

The theorem is powerful as it enables one to check global stability
from information only on the Jacobian matrix at the origin, provided
$0 \neq \| x \| \neq \| f(x) \|$ everywhere. As it stands, however, it says nothing
about characteristic roots of $J_f(x)$ for any x. There is, however, a
well-known theorem of Olech(1963) on global stability of the origin of
a system of two differential equations,

(2) $dx_i/dt = g_i(x_1, x_2)$,   i=1,2,

where $g = (g_1, g_2)'$: $R^2 \to R^2$, $g(0) = 0$, $g \in c^1$.
According to Olech, the origin is globally stable if characteristic
roots of the Jacobian matrix $J_g(x)$ have only negative real parts, that
is,

(3) $tr(J_g(x)) < 0$,

(4) $|J_g(x)| > 0$,

everywhere coupled with either

          $\partial g_1/\partial x_1 \partial g_2/\partial x_2 \neq 0$, or
(5)
          $\partial g_1/\partial x_2 \partial g_2/\partial x_1 \neq 0$

In the Olech theorem, global stability is closely related with the characteristic roots of $J_g(x)$, and a formal similarity between local and
global stability conditions is impressive as

          $tr(J_g(0)) < 0$ and $|J_g(0)| > 0$

constitute a necessary and sufficient condition for local stability.

In the light of these observations, we shall elaborate on the
Brock-Scheinkman theorem in such a way that its relationship with the
characteristic roots of the Jacobian matrix $J_f(x)$ or of $J_f(x)'J_f(x)$ will

become clear. To do so, define spectral norm of any complex matrix A which is not necessarily square by

$$Sp(A) \equiv (\mu(A^*A))^{1/2}$$

$$= \operatorname*{Sup}_{y \neq 0} \| Ay \| / \| y \| = \operatorname*{Sup}_{\| y \| = 1} \| Ay \| ,$$

where $\mu(.)$ means "the largest modulus of characteristic roots of" and y Euclidean norm of a vector y. See Lancaster(1969) and Pearce(1974). From the definition of a spectral norm, the first assumption in the Brock-Scheinkman theorem,

$$\| J_f(0)x \| < \| x \| , \quad x \neq 0$$

is equivalent to

(6) $$Sp(J_f(0)) = (\mu(J_f(0)'J_f(0))^{1/2} < 1.$$

If $J_f(0)$ is symmetric, (6) reduces to

$$\mu(J_f(0)) < 1$$

and $J_f(0)$ has characteristic roots whose moduli are all less than one, which corresponds to a valid local stability condition irrespective of the symmetricity of the Jacobian matrix evaluated at the origin. The second assumption in the Brock-Scheinkman theorem,

$$0 \neq \| x \| = \| f(x) \| \quad \text{implies} \quad \| J_f(x)x \| < \| x \|$$

is certainly met if

$$Sp(J_f(x)) = \operatorname*{Sup}_{y \neq 0} \| J_f(x)y \| / \| y \| < 1.$$

In the case of symmetric $J_f(x)$ this is rewritten as

$$\mu(J_f(x)) < 1.$$

Summarily, we have proved

Theorem 1.3.2.2: If $J_f(x)$ is symmetric and has characteristic roots
          whose moduli are all less than one <u>everywhere</u>, the origin of
          (1) is globally stable.

     We next consider a more general case not requiring symmetricity of the Jacobian matrix. As is easily seen, the origin is globally stable provided

$$Sp(J_f(x)) = (\mu(J_f(x)'J_f(x))^{1/2} < 1$$

everywhere, that is,

(7) $$\mu(J_f(x)'J_f(x)) < 1$$

everywhere. As $J_f(x)'J_f(x)$ is positive definite, its characteristic roots are all positive, provided $J_f(x)$ is non-singular. Thus (7) implies that

$$J_f(x)'J_f(x) - I$$

is negative definite as the matrix has only negative characteristic roots. In short,

Theorem 1.2.2.3: The origin of (1) is globally stable provided $J_f(x)'J_f(x) - I$ is negative definite.

Samuelson(1947) earlier has shown that a linear system of difference equations

$$x_{t+1} = Ax_t,$$

where A is a constant matrix, is stable if a matrix $A'A - I$ is negative definite. The correspondence between global stability Theorem 1.3.2.3 and Samuelson's local stability theorem is worthy of note.

Let us now give a fuller analysis of a two variable system of difference equations,

(8) $x_{1,t+1} = f_i(x_{1t}, x_{2t}),$ $i=1,2,$

where

$$f = (f_1, f_2)' \in c^1, \quad f(0) = 0.$$

For simplicity of notation let

$$J_f(x) = \begin{bmatrix} a_{11} & a_{12} \\ & \\ & \\ a_{21} & a_{22} \end{bmatrix} \equiv A,$$

where note should be made of a_{ij}'s dependence on time through x. The characteristic equation of $A'A$ is

$$|\lambda I - A'A| = \lambda^2 - (trA'A)\lambda + |A|^2 = 0$$

The condition for $\lambda_i < 1$ for i = 1, 2 consists of

$$(\lambda_1 - 1) + (\lambda_2 - 1) < 0$$

and

$$(\lambda_1 - 1)(\lambda_2 - 1) > 0.$$

We now claim

Theorem 1.3.2.4: The origin of (8) is globally stable provided

      (9) $\text{trA'A} < 2$

      (10) $|A|^2 - \text{trA'A} + 1 > 0$

are simultaneously satisfied.

    Inequalities (9) and (10) correspond to Olech's inequalities (3) and (4). If (9) holds, (10) is satisfied, provided $|A| > 1$ or $|A| < -1$.

CHAPTER 2

UNIQUENESS OF THE COURNOT OLIGOPOLY SOLUTION

Though the stability condition for the Cournot oligopoly solution
in a discrete (difference equation) system in the previous chapter ensured uniqueness of the solution, we assumed its uniqueness in our analysis of global stability in continuous (differential equation) systems.[1]

To see that the Cournot oligopoly solution is not necessarily unique,
consider a duopoly where two firms, A and B, produce identical goods
costlessly. See Fig. 3. The downward sloping curve DD' is the market
demand curve. If A's output in period t-1 is oX_1, B's demand curve in
period t becomes a downward sloping curve CD', B's iso-profit curves
are FF', GG' etc., as production is assumed to be costless. Two curves
CD' and FF' touch at E, corresponding to which B's profit maximizing
output in period t equals X_1X_2. Since A and B are interchangeable, A's
profit maximizing output in period t is also X_1X_2 provided B's output
in period t-1 is $oX_1 = X_1X_2$. Thus E is a Cournot oligopoly solution for
which both A and B's optimal outputs are oX_1. Arguing similarly, it may
be proved that E' corresponds to another solution where two firms' outputs are the same and equal $oX_3 = X_3X_4$.

Cournot conjectured that if the market demand function is decreasing
in price, the industry output in equilibrium will increase with an increase in the number of firms in the industry. Frank and Quandt(1963)
and McManus(1962, 1964) showed by examples that this conjecture lacks
general validity. McManus observed that under normal conditions on
demand function and with identical cost functions assumed for all firms,
the industry output for the Cournot oligopoly solution might be less
for a larger number of firms provided:

---

1) Note that under the stated conditions for Theorem 1.1.2.1 global
quasi-stability, instead of global stability, is ensured provided y*
(equilibrium or stationary point) is not unique. In Okuguchi(1964)
quasi-stability of the Cournot oligopoly solution was proved.

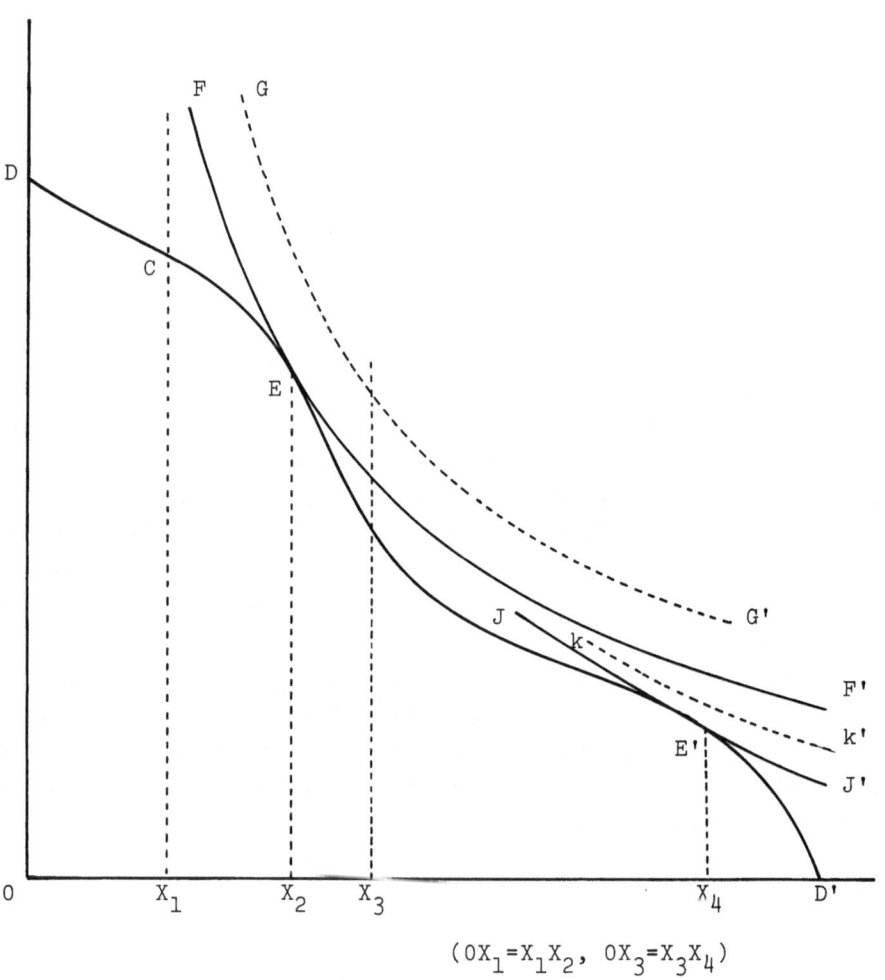

$(OX_1 = X_1X_2, \quad OX_3 = X_3X_4)$

Fig 3

(a) Marginal cost decreases.

(b) Marginal cost increases with multiple equilibria both before and after increase in the number of firms.

(c) Marginal cost is zero and the industry with a smaller number of firms has multiple equilibria.

In the light of these observations and for the necessity of assuming uniqueness in analyzing stability of the Cournot oligopoly solution, it is of much importance to investigate under what conditions the Cournot solution is unique. We strenghen Assumptions 1.1.1.2 and 1.1.2.3 to

Assumption 2.1: $p = f(\Sigma_i x_i)$ is twice continuously differentiable.

Assumption 2.2: $C_i = C_i(x_i)$ is twice continuously differentiable with $C_i(0) = 0$

Our result for the case of no product differentiation is stated as

Theorem 2.1: Under Assumptions 1.1.1.1, 1.1.1.4 - 1.1.1.5 and Assumptions1.1.2.1 - 1.1.2.2 and Assumptions 2.1 - 2.2, the (interior) Cournot oligopoly solution exists and is (globally) unique (or univalent).

Prior to proof of this theorem, mention must be made of

Theorem 2.2 (Gale and Nikaido(1965), Nikaido(1968): Let X be a closed rectangular region and $x = (x_1, \cdots, x_n) \in X$. Consider a mapping,

$$f(x) = (f^1(x), \cdots, f^n(x)): X \to R^n,$$

where f is differentiable on X. If the Jacobian matrix of f,

$$J_f(x) = [\partial f^i / \partial x_j]$$

has leading principal minors which are all positive for any $x \in X$ (that is, $J_f(x)$ is a P-matrix), then f(x) is univalent and

$$f(a) = f(b), \quad a, b \in X$$

is equivalent to a = b.

It has been proved by K.Inada(1971) that f(x) is univalent also if $J_f(x)$ has leading principal minors which are all negative for any $x \in X$ ($J_f(x)$ is a N-matrix). We are now ready to give

Proof of Theorem 2.1: Existence of the (interior) Cournot oligopoly solution under Assumptions 1.1.1.1, 1.1.1.4 - 1.1.1.5 and Assumptions 2.1 - 2.2 follows from Theorem 1.1.1.1. The solution x* satisfies the following equation.

$$f(\Sigma_j x_j^*) + x_i^* f'(\Sigma_j x_j^*) - C_i'(x_i^*) = 0, \qquad i=1,2,\cdots,n.$$

We define $g^1(x)$ by

$$g^1(x) \equiv -\{f(\sum_j x_j) + x_1 f'(\sum_j x_j) - C_1^!(x_1)\}, \quad x \in \Omega, \quad i=1,2,\cdots,n.$$

By virtue of Assumptions 2.1 and 2.2, $g^1(x)$ has a differential in the closed rectangular region Ω. The mapping,

$$G(x) = (g^1(x),\cdots,g^n(x)\ ), \quad \Omega \to R^n$$

is, therefore, differentiable for any $x \in \Omega$. Let $J_G(x)$ be the Jacobian matrix of G,

$$J_G(x) = [\partial g^1/\partial x_j],$$

where

$$\partial g^1/\partial x_1 \equiv g_1^1 = -\{2f'(\sum_k x_k) + x_1 f''(\sum_k x_k) - C_1''(x_1)\}, \quad i=1,2,\cdots,n$$

$$\partial g^1/\partial x_j \equiv g_j^1 = -\{f'(\sum_k x_k) + x_1 f'(\sum_k x_k)\}, \quad i \neq j, \quad i=1,2,\cdots,n.$$

We derive:

$$g_1^1 = -(-f' + C_1'') - (f' + x_1 f''),$$

$$\begin{vmatrix} g_1^1 & g_j^1 \\ g_1^j & g_j^j \end{vmatrix} = (-f' + C_1'')(-f' + C_j'') - (f' + x_1 f'')(-f' + C_j'')$$

$$- (f' + x_j f'')(-f' + C_1''), \quad i \neq j$$

$$\begin{vmatrix} g_1^1 & g_j^1 & g_k^1 \\ g_1^j & g_j^j & g_k^j \\ g_1^k & g_j^k & g_k^k \end{vmatrix} = (-f' + C_1'')(-f' + C_j'')(-f' + C_k'')$$

$$- (f' + x_1 f'')(-f' + C_j'')(-f' + C_k'')$$

$$- (f' + x_j f'')(-f' + C_1'')(-f' + C_k'')$$

$$- (f' + x_k f'')(-f' + C_1'')(-f' + C_j''),$$

$$i \neq j \neq k$$

$$|J_G(x)| = \prod_{i=1}^{n} (-f' + C_1'') - \sum_{i=1}^{n} (f' + x_1 f'') \prod_{j \neq i} (-f' + C_j'').$$

Due to Assumptions 1.1.2.1 and 1.1.2.2, all leading principal minors of the Jacobian matrix $J_G(x)$ are thus seen to have positive signs for any $x \in \Omega$. Finally, we apply Theorem 2.2 to conclude that the equation

$$G(x) = 0, \quad x \in \Omega$$

has a unique solution. Hence uniqueness of the Cournot oligopoly solution.   Q.E.D.

CHAPTER 3

ENTRY IN THE COURNOT MODEL:

QUASI-COMPETITIVENESS VS PERFECT COMPETITION

3.1. Introductory Remarks

As might be easily conjectured, the Cournot oligopoly solution and perfect competitive equilibrium generally differ. To see this by way of an example due to Mayberry, Nash and Shubik(1953)(see also Shubik (1959)), let two firms' cost functions in a Cournot duopoly under no product differentiation be given by

$$C_1(x_1) = 4 - x_1 + x_1^2,$$

$$C_2(x_2) = 5 - x_2 + x_2^2,$$

and let

$$p = 10 - 2(x_1 + x_2)$$

be the market demand function. If two firms act in perfect competition, marginal costs of both must equal the competitively given market price. Solving

$$\partial C_1/\partial x_1 = \partial C_2/\partial x_2 = p,$$

$$x_1 = 1.1716, \quad x_2 = 0.9411, \quad p = 5.7747$$

are obtained. Under the Cournot assumption,

$$\partial \pi_1/\partial x_1 = \partial \pi_2/\partial x_2 = 0$$

must hold in equilibrium. Thus,

$$x_1 = 0.9386, \quad x_2 = 0.7400, \quad p = 6.6428.$$

We may note that two firms' outputs are both larger and the market price smaller in perfect competition than in the Cournot equilibrium.

Two important problems arise regarding entry in the Cournot model. The first is quasi-competitiveness of the Cournot oligopoly solution and concerns whether the industry output for the Cournot equilibrium will increase or not with an increase in the number of firms in the industry. The second concerns convergence of the Cournot oligopoly solution to perfect competitive equilibrium as the number of firms goes to infinity. These two problems were first taken up by C.R.Frank Jr.,

(1965). R.J.Ruffin(1971) also analyzed the same problems for a <u>strong</u>
<u>case</u> where all firms in the industry are assumed to have identical cost
functions. I.Horowitz(1970) approached quasi-competitiveness via a
simple example. As was mentioned in Chapter 2, McManus(1964) noticed
a close relationship between uniqueness of the Cournot oligopoly solution and quasi-competitiveness. The uniqueness problem was analyzed
systemmatically in the same chapter.

In this chapter we shall be concerned with quasi-competitiveness
of the Cournot oligopoly solution and its possibility of convergence
to perfect competitive equilibrium for <u>a weak case</u> where differences
in cost functions among firms are assumed to exist. Though Frank(1965)
in his proof of convergence of the Cournot oligopoly solution to competitive equilibrium took explicitly into consideration differences in
cost functions among firms, his approach differs radically from ours
in that we assume initial existence of an equal number of firms for
each type of conceivable cost function and let the number increase
equally and simultaneously for each type of cost function. Our approach
is in parallel to that of G.Debreu and H.Scarf(1963). In fact, they
proved equivalence of the core and competitive equilibrium in a pure
exchange economy for which they assumed existence of an equal number
of exchage participants for each category of initial endowment and the
participants were then increased equally and successively in each
category. We shall find that the conditions for uniqueness of the
Cournot oligopoly solution are sufficient for quasi-competitiveness.

3.2.Quasi-Competitiveness

We assume that there exist m types of cost functions and N firms
for each type, the total number of firms thus being equal to mN.
Necessary notation and assumptions are as follows:

$F_{(i, j)}$; the j-th firm having cost function of the i-th type,

x_i^j, output of $F_{(i, j)}$,

$C_i^j(x_i^j) = C_i(x_i^j)$, cost function of $F_{(i, j)}$,

$Q \equiv \sum_{i,j} x_i^j$, industry output,

p = f(Q), market demand function, where p is the market price of a
homogeneous goods.

Subscripts and superscripts range over 1 through m and 1 through N,
respectively.

<u>Assumption 3.2.1</u>: $x_i^j \in \Omega_i = [0, M_i]$, where M_i is a finite positive

number.

Assumtpion 3.2.2: $p = f(Q) \in C^2$.

Assumption 3.2.3: $C_i(x_i^j) \in C^2$ on Ω_i, $C_i(0) = 0$ and $C_i(x_i^j)$ is strictly
          increasing in x_i^j.

Assumption 3.2.4: Profit function of $F_{(i, j)}$,

$$\pi_I^j = x_i^j f(Q) - C_i(x_i^j)$$

is concave in x_i^j.

Ruffin(1971) assumes strict concavity of profit function for each firm.
Note, however, that Assumption 3.2.6 and 3.2.7 below imply strict
concavity.

Assumption 3.2.5:

$$C_i'(0) < f(\sum_{k,l} x_k^l - x_i^j),$$

$$f(M_i + \sum_{k,l} x_k^l - x_i^j) + M_i f'(M_i + \sum_{k,l} x_k^l - x_i^j) < C_i'(M_i).$$

This assumption is needed to ensure existence of the interior Cournot
solution.

Assumption 3.2.6: $f' + x_i^j f'' < 0$.

This assumption says that marginal revenue of $F_{(i, j)}$(with respect to
change in its own output), $f + x_i^j f'$, is a strictly decreasing function
of any other firm's output. Put differently, the marginal revenue
curve of $F_{(i, j)}$ has a steeper slope than the market demand function.

Assumption 3.2.7: $f' < C_i''(x_i^j)$.

This assumption is satisfied if the demand curve is downward sloping and
the marginal cost of each firm is either constant or increasing. It is
also met when marginal costs of all firms are decreasing at slower rates
than the rate of change in market price.

    We now apply Theorem 2.1 to assert that the unique interior Cournot
oligopoly solution exists under Assumption 3.2.1 - 3.2.7. For the
Cournot solution,

(1) $x_i^{j*} f'(\sum_{i,j} x_i^{j*}) + f(\sum_{i,j} x_i^{j*}) - C_i'(x_i^{j*}) = 0,$ $i=1,\cdots,m,$
                                                                                            $j=1,\cdots,N,$

As the solution is unique, it follows that x_i^{j*}, equilibrium output of
$F_{(i, j)}$, is identical for all firms having identical cost functions.

Let $x_i^{j*} = x_i^*$ to rewrite (1) as

(2) $x_i^* f'^* + f^* - C_i'^* = 0,$ $i=1,2,\cdots,m,$ $j=1,2,\cdots,N,$

where, for notational simplification, f^* and f'^* denote f and f'

evaluated at $Q = Q^* = N \sum_{j=1}^{m} x_j^*$, and $C_i^!*$ stands for $C_i^!$ at $x_i^j = x_i^*$.

Solving (2), x_i^* can be expressed as a function of N,

(3) $x_i^* = x_i^*(N)$, $i=1,2,\cdots,m$.

The industry output in equilibrium is also a function of N,

(4) $Q^* = Q^*(N) = N \sum_{j=1}^{m} x_j^*(N)$.

To see the effects of increase in the number of firms, totally differentiate (2) to get,

$$(f'^* - C_i'')dx_i^*/dN + (x_i^*f''^* + f'^*)N\Sigma_j dx_j^*/dN$$

$$= -(x_i^*f''^* + f'^*)Q^*/N, \qquad i=1,2,\cdots,m,$$

or in matrix notation,

(5)
$$
\begin{bmatrix}
b_1 & a_1 & a_1 & \cdots & a_1 \\
a_2 & b_2 & a_2 & \cdots & a_2 \\
\vdots & & \ddots & & \vdots \\
a_m & a_m & a_m & \cdots & b_m
\end{bmatrix}
\begin{bmatrix}
dx_1^*/dN \\
dx_2^*/dN \\
\vdots \\
dx_m^*/dN
\end{bmatrix}
= -Q^*/N
\begin{bmatrix}
a_1 \\
a_2 \\
\vdots \\
a_m
\end{bmatrix}
$$

where

$$a_i \equiv (x_i^*f'' + f'^*)N, \qquad i=1,2,\cdots,m$$

$$b_i \equiv (f'^* - C_i''^*) + (x_i^*f''^* + f'^*)N, \qquad i=1,2,\cdots,m.$$

Let A be the coefficient matrix of the left-hand side of (5). A rather complicated manipulation enables one to derive,

(6) $$|A| = \prod_{j=1}^{m}(f'^* - C_j''^*) + \sum_{i=1}^{m}(x_i^*f''^* + f'^*)N \prod_{j \neq i}^{m}(f'^* - C_j''^*)$$

From Assumption 3.2.6 and 3.2.7, $|A|$ is seen to be positive or negative according to the evenness or oddness of m. Solving (5),

(7) $$dx_i^*/dN = -Q^*(x_i^*f''^* + f'^*) \prod_{j \neq i}^{m}(f'^* - C_j''^*)/N|A|, \qquad i=1,\cdots,m.$$

Due to Assumptions 3.2.6 and 3.2.7, the numerator (without the minus sign) of the right-hand side of (7) is positive or negative according to the evenness or oddness of m. Hence,

(8) $dx_i^*/dN < 0$, $i=1,2,\cdots,m$

regardless of whether m is odd or even, proving that a simultaneous

increase in number of firms having different types of cost functions
is conducive to reduction in each firm's output in equilibrium.

In order to investigate the effects of entry or increase in N on
the industry output in equilibrium, differentiate (4) with respect to
N and substitute (7) to derive,

$$(9) \qquad dQ*/dN = \sum_{j=1}^{m} x_j^* + N \sum_{j=1}^{m} dx_j^*/dN$$

$$= Q* \prod_{j=1}^{m} (f'^* - C_j''^*)/N|A|.$$

This expression is unambiguously positive, showing that increase in N
necessitates increase in the industry output corresponding to the
Cournot oligopoly solution. Hence,

Proposition 3.2.1: The Cournot oligopoly solution is quasi-competitive
          under our assumptions and the industry output in equilibrium
          increases, hence equilibrium market price decreases, with
          simultaneous increase in number of firms in each category of
          cost function.

Note, however, that no definite statement is possible regarding the
effects of increase in N on the equilibrium industry output when
Assumption 3.1.2.7 does not hold for at least one firm, as the sign of
the last expression of (9) becomes indeterminate in this case.

3.3. Convergence to Perfect Competition

We have to introduce the following assumptions for establishment
of convergence of the Cournot oligopoly solution to perfect competitive
equilibrium.

Assumption 3.3.1: There exists a finite positive number G such that
          $|f'(Q)| \leq G$.

Assumption 3.3.2: A finite positive number $\bar{Q}$ exists so that $f(Q) = 0$
          for $Q \equiv \sum_{i,j} x_i^j \geq \bar{Q}$.

Assumption 3.3.2 means that homogeneous goods produced by all firms
become free whenever their total supply is equal to or greater than a
finite amount $\bar{Q}$. This assumption coupled with Assumption 3.2.3 on cost
functions leads to

$$Q* = \sum_{i,j} x_i^{j*} = N\sum_{i} x_i^* \leq \bar{Q} \qquad \text{for all N,}$$

from which $x_i^* \leq \bar{Q}/N$. Thus

$$x_i^* \to 0 \text{ as } N \to \infty \qquad \text{for all i.}$$

This fact together with Assumption 3.3.1 entails

$x_i^* f'^* \to 0$ as $N \to \infty$ for all i.

Expression (2) in Section 3.2 thus reduces to

(1) $\qquad f^* = C_i'^*, \qquad i=1.2. \cdots, m,$

which shows that the market price equals marginal costs of all firms
and $x_i^* = 0$ for all i when N goes to infinity. To proceed further,
define average cost of $F_{(i, j)}$ as

$$AC_i^j(x_i^j) \equiv C_i^j(x_i^j)/x_i^j = C_i(x_i^j)/x_i^j.$$

Take into account $C_i^j(0) = 0$ for all i and j, and apply L'Hospital's rule
to derive,

(2) $\qquad \lim_{x_i^j \to 0} A_i^j(x_i^j) = \lim_{x_i^j \to 0} C_i'/1 = C_i'(0).$

Thus marginal and average costs take the same value for all firms as N
goes to infinity. In (long-run) perfect competitive equilibrium, price
equals marginal cost equals minimum average cost. Hence the following
propositions.

Proposition 3.3.1: Under Assumption 3.3.1 and 3.3.2, the Cournot
oligopoly solution converges to long-run perfect competitive
equilibrium provided $C_i'(0) = \min_{x_i^j} A_i^j(x_i^j)$ for all i and j.[1]

Proposition 3.3.2: If $C_i(x_i^j)$ is U-shaped with $C_i'(0) > \min_{x_i^j} A_i^j(x_i^j)$ for

at least one i (and j), the alluded convergence will not
occur.[2]

To illustrate our results, let us consider a simple example[3] of
m = 1 and (suppress suffix to C)

(3)
$$C(x^j) = \alpha x^j, \quad \alpha > 0, \quad j=1,2,\cdots,N$$

$$p = a - b\Sigma_j x^j, \quad a, b > 0 \quad \text{for } Q = \Sigma_j x^j \le a/b = \bar{Q}$$

$$= 0 \qquad\qquad\qquad \text{for } Q \ge \bar{Q}.$$

The Cournot oligopoly solution satisfies:

$$p^* + x^{j^*} f'^* - C'(x^{j^*}) = 0, \qquad j=1,2,\cdots,N.$$

---

1) This proposition corresponds to Ruffin(1971, Theorem 1).
2) This proposition corresponds to Ruffin(1971, Theorem 2).
3) This example is suggested by I.Horowits(1970).

Solving this and letting $x^{j*} = x*$ for all j, we have:

$$x* = (a - \alpha)/(N + 1)b, \quad (\text{Assume } a > \beta)$$

$$Q* = Nx* = N(a - \beta)/(N + 1)b,$$

$$p* = (a + N\beta)/(N + 1),$$

$$\pi^{j*} \equiv \pi* = (a - \beta)^2/(N + 1)^2\beta, \qquad j=1,2,\cdots,N.$$

Partially differentiating these expressions with respect to N:

$$\partial x*/\partial N < 0, \ \partial Q*/\partial N > 0, \ \partial p*/\partial N < 0, \ \partial \pi*/\partial N < 0.$$

Taking limits as N goes to infinity:

$$\lim_{N\to\infty} x* = 0, \ \lim_{N\to\infty} Q* = (a - \beta)/b < a/b = \bar{Q}$$

$$\lim_{N\to\infty} p* = \beta, \ \lim_{N\to\infty} \pi* = 0.$$

These calculations show that our simple example for which there exists only one type of cost function has the property of quasi-competitiveness. Further, as

$$\beta = \lim_{N\to\infty} p* = \underset{x*\to0}{MC} (x*) = \underset{x*\to0}{AC} (x*) = \min_{x_i^j} AC(x^j) \text{ for all } j,$$

convergence of the Cournot oligopoly solution to perfect competitive equilibrium occurs without fail. Note that for our example

$$G = b < \infty, \ \bar{Q} = a/b < \infty, \ C(0) = 0$$

and $C(x^j)$ is strictly increasing with $MC = AC = \beta$ for all x^j.

Though we have analyzed the effects of entry in a Cournot model, we have not given any analysis of entry mechanism. The reader is referred to M.I.Kamien and N.L.Schwartz(1975) for an explicit analysis of this.

CHAPTER 4

REVENUE MAXIMIZING DUOPOLY

4.1. Introduction

Oligopolists in our analyses so far were unconstrained profit
maximizers. W.J.Baumol(1958, 1959) maintains that revenue (or sales)
maximization under a constraint of attainment of minimum necessary
profit is more commonly observed among large firms than unconstrained
profit maximization, and has formulated a model of revenue maximization
under minimum profit constraint. Though Baumol is well aware of interdependence among firms, he has ignored this in his analysis. To cite
from his book: "I shall take the position that in _day_ _to_ _day_ _decision_
making, management often acts explicitly or implicitly on the premise
that its decisions will produce no changes in the behavior of those
with whom they are competing. Of course, I am not arguing that management inhabits a fool's paradise in which interdependence is never considered. In making its more radical decisions, such as launching of
a major advertizing campaign or the introduction of a radically new
line of products, management usually does not consider the probable
competitive response".[1]

W.G.Shepherd(1962) and C.J.Hawkins(1970) have pointed out the importance of interdependence among firms in a constrained revenue maximization oligopoly, but a successful integration of mutual interdependence of firms and constrained revenue maximization hypothesis into a
model was first made by J.F.Formby(1973). He considered a simple case
of duopoly under no product differentiation and assumed that each duopolist had identical and constant average cost = marginal cost and that
market demand function was linear.

As is easily imagined, different models of constrained revenue
maximizing duopoly (or more generally oligopoly) will emerge depending
on differences   regarding hypothesis on mutual interdependence among
firms.

As for a constrained revenue maximizing duopoly, five possibilities
arise. To wit:

(a) Two firms are both unconstrained revenue maximizers for which
minimum profit constraints are not binding.

---

1) Baumol(1959, pp. 28-29). Italics is his.

(b) One firm is an unconstrained revenue maximizer and the other realizes just minimum required profit.

(c) Both firms are realizing only minimum required profits.

(d) One firm is an unconstrained revenue maximizer, while the other one is striving for maximum profit which is less than the minimum required level.

(e) Two firms are both profit maximizers unable to attain the minimum required level.

In order to conduct a diagrammatical analysis of stability of the Cournot duopoly solution under constrained revenue maximization, constrained reaction function must be derived. Consider a simple case where

$$P = a - b(x_1 + x_2), \quad C_1 = c_1 x_1, \qquad i=1,2.$$

Under the Cournot assumption of the rival's behavior, reaction function for unconstrained profit maximization of the i-th firm, $h_i(x_j(t-1))$, is obtained by maximizing its expected profit in period t,

$$x_i(t)\{a - b(x_i(t) + x_j(t-1))\} - c_i x_i(t).$$

(1) $$x_i(t) = h_i(x_j(t-1)) = -x_j(t-1)/2 + (a - c_i)/2b,$$

$$i \neq j, \quad i,j = 1,2.$$

Suppose, on the other hand, that the i-th firm maximizes its revenue under the Cournot assumption and without any constraint on its profit, then,

(2) $$x_i(t) = -x_j(t-1)/2 + a/2b, \qquad i \neq j, \quad j=1,2.$$

Satisfaction of minimum profit constraint under the Cournot assumption requires

(3) $$(a - c_1)x_1(t) - bx_1(t)^2 - bx_1(t)x_j(t-1) \geq \bar{\pi}_1,$$

$$i \neq j, i,j=1,2,$$

where $\bar{\pi}_i$ is the minimum required profit level of the i-th firm.

On the understanding that x_1 and x_2 refer to period t and period t-1, respectively, the constrained reaction function for the firm 1 which strives for attainment of maximum revenue under minimum profit constraint is depicted by a real curve αδδ'α' in Fig 4, where the shaded region below a curve γδδ'γ' defined by

$$x_2 = (a - c_1)/b - x_1 - \bar{\pi}_1/b_1 x_1$$

satisfies minimum profit constraint; a line passing through αδα' expresses

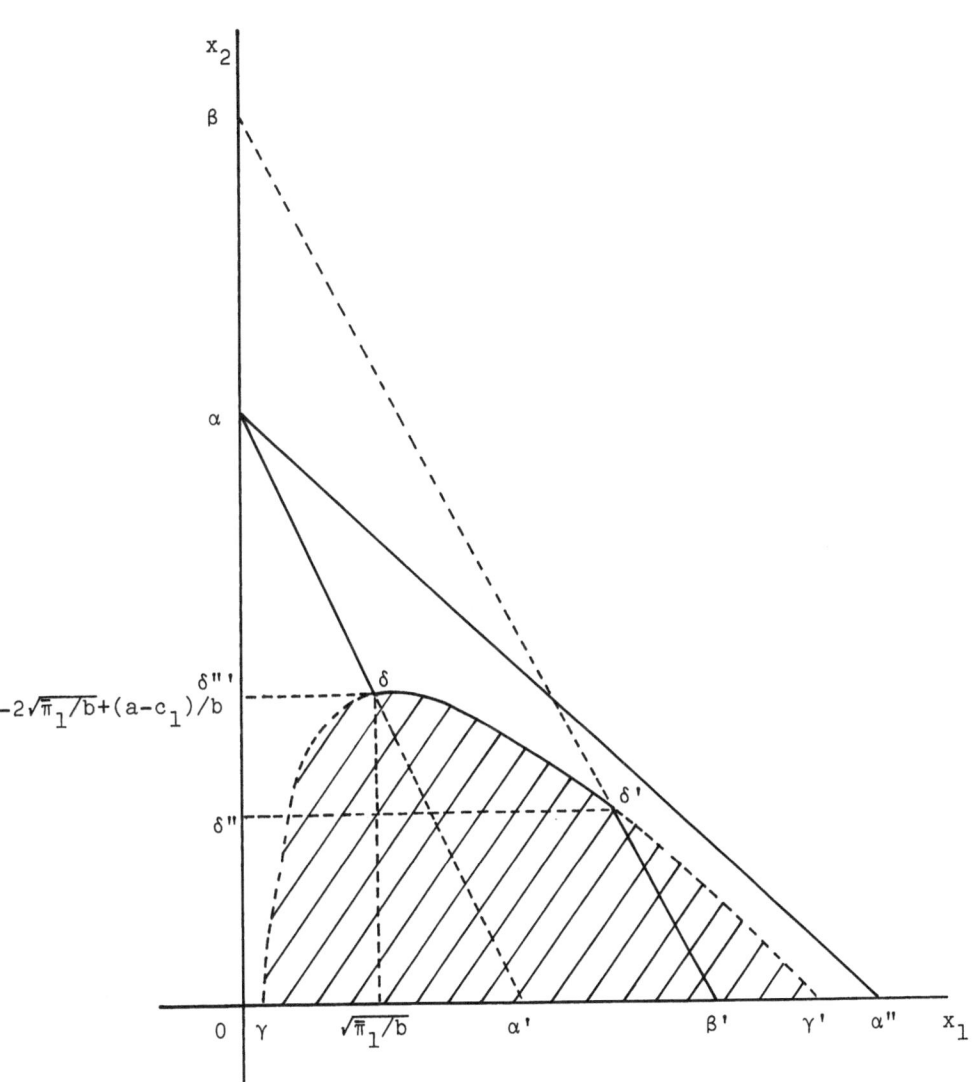

Fig 4

$$x_1 = -x_2/2 + (a - c_1)/2b;$$

a line passing through $\beta\delta'\beta'$ corresponds to

$$x_1 = -x_2/2 + a/2b;$$

and a line through $\alpha\alpha''$ denotes $x_2 = (a - c_1)/b - x_1$, one of the asymptotes to the curve $\gamma\delta\delta'\gamma'$. Note that Fig 4 is depicted on an implicit understanding that a line through $\beta\beta'$ intersects the curve $\gamma\delta\gamma'$ at δ'. If c_1 is too large, δ' does not exist and the constrained reaction function consists of a line $\alpha\delta$ and a curve $\delta\delta'\gamma'$. Constrained reaction function for firm 2 may similarly be derived. For firm 2, however, x_2 refers to period t and x_1 to period t-1.

The Cournot duopoly solution under constrained revenue maximization, if it ever exists and if it is unique or not, is given by the intersection of two firms' constrained reaction functions. So let

$$x_i = g_i(x_j), \qquad i \neq j, \ i,j = 1,2$$

be the i-th firm's constrained reaction function, and x_i^* its output in equilibrium. Formby showed, diagrammatically, that the Cournot duopoly solution is locally stable if and only if the slope evaluated at the equilibrium of firm 1's reaction function is steeper than that for firm 2, that is,

(4) $$1/|g_1'^*| > |g_2'^*|,$$

where the asterisk means evaluation at equilibrium. This stability condition is <u>formally</u> identical to that for an unconstrained Cournot duopoly with instantaneous adjustment.

4.2. Stability Analysis

In this section we shall try an analytical approach to stability of the constrained revenue maximizing Cournot duopoly solution in a general setting where market demand function and two firms' cost functions are not restricted to be linear. Reaction function in constrained revenue maximizing duopoly as well as in unconstrained profit maximizing Cournot duopoly reflects properties of demand, revenue and cost functions. Our stability condition will be given in such a form that this point is well taken care of. We shall use a discrete model. Let

$$p = f(x_1 + x_2), \quad f' < 0$$

be market demand function and $C_i = C_i(x_i)$, $i = 1,2$, be the i-th firm's cost function. We assume differentiability of relevant functions as many times as required. From what was mentioned in the previous section,

it is necessary to analyze stability of constrained revenue maximizing
Cournot duopoly solution under no product differentiation in five possible cases, (a-e).

(a) The case where two firms are both unconstrained revenue
maximizers.

Since under the Cournot assumption each firm assumes that its
rival's output in period t will be the same as in the previous one,
t-1, expected revenue of the i-th firm in period t is given as

(5) $\qquad R^1(x_i(t), x_j(t-1)) = x_i(t)f(x_i(t) + x_j(t-1)), \quad i \neq j, \ i,j=1,2.$

Maximization of this with respect to $x_i(t)$ yields,

(6) $\qquad f(x_i(t) + x_j(t-1)) + x_i(t)f'(x_i(t) + x_j(t-1)) = 0.$

$$i \neq j, \ i,j=1,2.$$

Solving this to derive the reaction function,

(7) $\qquad x_i(t) \equiv g_i(x_j(t-1)), \quad i \neq j, \ i,j=1,2,$

where by total differentiation of (6),

(8) $\qquad dx_i(t)/dx_j(t-1) = -\{x_i(t)f''(x_i(t) + x_j(t-1))$

$\qquad\qquad + f'(x_i(t) + x_j(t-1))\}/\{x_i(t)f''(x_i(t) + x_j(t-1))$

$\qquad\qquad + 2f'(x_i(t) + x_j(t-1))\}, \qquad i \neq j, \ i,j=1,2.$

Define:

$\qquad MR^1 \equiv \partial R^1/\partial x_1,$

(9) $\qquad \partial MR^1/\partial x_i(t) \equiv MR^1_i = x_i(t)f''(x_i(t) + x_j(t-1)) + 2f'(x_i(t)$

$\qquad\qquad +x_j(t-1)), \qquad i \neq j, \ i,j=1,2.$

(10) $\qquad \partial^2 R^1/\partial x_i(t)\partial x_j(t-1) \equiv MR^1_j = x_i(t)f''(x_i(t) + x_j(t-1))$

$\qquad\qquad + f'(x_i(t) + x_j(t-1))$

$$i \neq j, \ i,j=1,2.$$

Expressions (9) and (10) are nothing but the denominator and numerator of (8), respectively. From (7),

$\qquad x_i^* = g_i(x_j^*), \qquad i \neq j, \ i,j=1,2.$

Noting that x_2^* is stable if and only if x_1^* is so, define $X_1 \equiv x_1 - x_1^*$.
Taking into account the equilibrium conditon and applying Taylor ex-
pansion to (7), the following difference equation which is fundamental

to stability analysis is obtained.

(11) $\qquad X_1(t) = g_1'*g_2'*X_1(t-1).$

The equilibrium is thus stable if and only if

(12) $\qquad |g_1'*g_2'*| = |MR_2^{1*}|/|MR_1^{1*}| \cdot |MR_1^{2*}|/|MR_2^{2*}| < 1.$

where the asterisk means that derivatives are evaluated at equilibrium.
The stability conditions are satisfied provided

$$|MR_2^{1*}| / |MR_1^{1*}| < 1, \quad |MR_1^{2*}|/|MR_2^{2*}| < 1$$

are simultaneously fulfilled. Thus stability is ensured if the absolute
value of each firm's rate of change of expected marginal revenue with
respect to change in its own output is greater than that with respect
to change in its rival's output, which is true if $MR_2^{1*} < 0$ and $MR_1^{2*} < 0$
as $f'* < 0$. Note that in this case of unconstrained revenue maximization, each firm's cost function has nothing to do with stability.

(b) One firm is an unconstrained revenue maximizer and the other
one realizes just a minimum required profit level.

We may let, without loss of generality, firm 1 be the unconstrained
revenue maximizer and firm 2 attain only minimum profit. As in (a),
unconstrained revenue maximizing condition for firm 1 is,

(13) $\qquad f(x_1(t) + x_2(t-1)) + x_1(t)f'(x_1(t) + x_2(t-1)) = 0.$

For firm 2, attainment of only minimum profit level entails,

(14) $\qquad x_2(t)f(x_1(t-1) + x_2(t)) - C_2(x_2(t)) = \bar{\pi}_2.$

Equations (15) and (16) derivable from (13) and (14), respectively,
are reaction functions for firm 1 and firm 2, respectively.

(15) $\qquad x_1(t) = g_1(x_2(t-1)),$

(16) $\qquad x_2(t) - g_2(x_1(t-1)),$

where

(17) $\qquad dx_1(t)/dx_2(t-1) = -MR_2^1/MR_1^1,$

(18) $\qquad dx_2(t)/dx_1(t-1) = -x_2(t)f'(x_1(t-1) + x_2(t))/\{f(x_1(t-1)$

$$+ x_2(t)) + x_2(t)f'(x_1(t-1) + x_2(t))$$

$$- C_2'(x_2(t))\}$$

$$= -\partial R^2/\partial x_1(t-1)/(MR^2 - MC_2),$$

and MC_2 stands for firm 2's marginal cost. Arguing similarly as in (a), stability is shown to be ensured if and only if $|g_1'*g_2'*| < 1$, that is,

(19) $|MR_2^{1*}|/|MR_1^{1*}| \cdot |(\partial R^2/\partial x_1)*|/|MR^{2*} - MC_2^*| < 1.$

(c) Each firm just attains minimum profit level.

In this case, a slight adaptation of arguments in (b) enables one to assert that a necessary and sufficient condition for stability is

(20) $|(\partial R^1/\partial x_2)*|/|MR^{1*} - MC_1^*| \cdot |(\partial R^2/\partial x_1)*|/|MR^{2*} - MC_2^*| < 1.$

(d) Firm 1 is an unconstrained revenue maximizer and firm 2 makes the best of a bad situation and maximizes profit which is less than the minimum required level.

In this case the following equations have to hold,

$$\partial R^1(x_1(t), x_2(t-1))/\partial x_1(t) = 0, \quad \partial \pi_2(x_1(t-1), x_2(t))/\partial x_2(t) = 0,$$

where we define,

$$\pi_2(x_1(t-1), x_2(t)) \equiv x_2(t)f(x_1(t-1) + x_2(t)) - C_2(x_2(t)).$$

Hence, $dx_1(t)/dx_2(t-1) = -MR_2^1/MR_1^1,$

$$dx_2(t)/dx_1(t-1) = -MR_1^2/(MR_2^2 - MC_2').$$

Stability follows if and only if

(21) $|MR_2^{1*}|/|MR_1^{1*}| \cdot |MR_1^{2*}|/|MR_2^{2*} - MC_2'*| < 1.$

(e) Each firm makes the best of a bad situation and maximizes profit which can not attain the minimum required level.

This case corresponds to unconstrained profit maximizing Cournot duopoly under no product differentiation where adjustment is instantaneous. Similar arguments as for firm 2 in (d) for both firms lead to the following stability condition.

(22) $|MR_2^{1*}|/|MR_1^{1*} - MC_1'*| \cdot |MR_1^{2*}|/|MR_2^{2*} - MC_2'*| < 1.$

This inequality holds provided

$$MR_j^{1*} < 0, \quad f'* < MC_i'*, \quad i \neq j, \ i,j=1,2.$$

CHAPTER 5

STACKELBERG DUOPOLY MODELS RECONSIDERED

5.1. A Leader-Follower Model

In spite of the extensive works done on problems of the Cournot
(or Cournot-type) oligopoly models, dynamic analysis of'duopoly model
formulated by H.von Stackelberg(1934) has almost been outside the scope
of concern of economists. J.M.Henderson and R.E.Quandt(1958) in now a
classic on mathematical microeconomics have given an excellent summary
of Stackelberg's fundamental idea on duopoly without, however, giving
any systematic analysis of dynamics. As is well known, three cases,
that is, follower-follower, leader follower and leader-leader, are
considered by Stackelberg. Since any duopolist is in an advantageous
position when he becomes a leader, both duopolists strive for leadership. However, "If both desire to be leaders, each assumes that the
other's behavior is governed by his reaction function, but in fact,
neither of the reaction function is observed and a Stackelberg disequilibrium is encountered. Stackelberg believed that this disequilibrium is the most frequent outcome. The final result of a Stackelberg
disequilibrium cannot be predicted a priori. If Stackelberg was correct,
this situation will result in economic warfare, and equilibrium will
not be achieved until one has succumbed to the leadership of other or
a collusive agreement has been reached". (Henderson and Quandt(1958,
p. 181))

We owe to D.R.Kamerschen and P.Smith(1971) the first dynamic analysis of Stackelberg models of duopoly under no product differentiation.
Since the leader-leader case where each duopolist assumes that the
other is a follower and is satisfied himself with being a follower,
coincides with the Cournot model, we have to consider only the leader-
follower and leader-leader cases. As Henderson and Quandt observed,
the ultimate result in the leader-leader case is indeterminate, which
motivates our analysis aiming at resolution of indeterminacy in the
leader-leader model in the following section.

In this section, we focus our attention on (existence and) stability of the leader-follower solution in a discrete model. Kamerschen
and Smith, however, deal with a continuous system. Let, without loss
of generality, firm 1 be the leader. Firm i's profit function and market demand function are,

(1) $\quad \pi_1(x_1, x_j) = px_1 - C_1(x_1), \qquad i \neq j, \ i,j=1,2,$

(2) $\quad p = f(x_1 + x_2),$

where p, x_1 and C_1 have the same meaning as in the previous chapters. In the following analysis differentiability of relevant functions up to required orders is assumed. Since firm 2 is the follower, it will choose its optimal output in period t to maximize its expected profit in the same period as given by

$$x_2(t)f(x_1(t-1) + x_2(t)) - C_2(x_2(t)).$$

The first order condition for expected profit maximization is (under an implicit understanding that a corner maximum is excluded),

(3) $\quad f(x_1(t-1) + x_2(t)) + x_2(t)f'(x_1(t-1) + x_2(t)) - C_2'(x_2(t)) = 0.$

This (under the assumption of the uniqueness of the maximizer) gives rise to

(4) $\quad x_2(t) = \phi_2(x_1(t-1)).$

The leader's optimal output in period t will maximize its expected profit in the same period. Taking into account (4), the first and second order conditions for firm 1's profit maximization are,

(5) $\quad f(x_1(t) + \phi_2(x_1(t-1))) + x_1(t)f'(x_1(t) + \phi_2(x_1(t-1)))$

$\qquad - C_1'(x_1(t)) = 0,$

(6) $\quad 2f'(x_1(t) + \phi_2(x_1(t-1))) + x_1(t)f''(x_1(t) + \phi_2(x_1(t-1)))$

$\qquad - C_1''(x_1(t)) < 0.$

From (5),

(7) $\quad x_1(t) = h_1(x_1(t-1)),$

(8) $\quad h_1' \equiv dx_1(t)/dx_1(t-1) = dx_1(t)/dx_2(t) \cdot dx_2(t)/dx_1(t-1),$

(9) $\quad dx_1(t)/dx_2(t) = -\{f'(x_1(t) + x_2(t)) + x_1(t)f''(x_1(t) + x_2(t))\}/$

$\qquad \{2f'(x_1(t) + x_2(t)) + x_1(t)f''(x_1(t) + x_2(t))$

$\qquad - C_1''(x_1(t))\},$

$$(10) \quad dx_2(t)/dx_1(t-1) = -\{f'(x_1(t-1) + x_2(t)) + x_2(t)f''(x_1(t-1)$$
$$+ x_2(t))/\{2f'(x_1(t-1)) + x_2(t))$$
$$+ x_2(t)f''(x_1(t-1) + x_2(t)) - C_2''(x_2(t))\}.$$

For any x_1 and x_2 in their respective domains, the following two assumptions are assumed to be satisfied.

Assumption 5.1.1: $f'(x_1 + x_2) + x_i f''(x_1 + x_2) < 0$, $i=1,2$.

Assumption 5.1.2: $f'(x_1 + x_2) < C_i''(x_1)$, $i=1,2$.

These assumptions state that marginal revenue of any firm is decreasing in another firm's output and the rate of change of each firm's marginal cost is algebraically greater than the rate of change of the market price with respect to change in total supply. We are now ready to prove

Theorem 5.1.1:   Under Assumptions 5.1.1 and 5.1.2, the leader-follower model under no product differentiation has a unique and globally stable solution (or equilibrium).

Proof: Applying the mean value theorem to (7), we have

$$|x_1(t) - x_1(t-1)| = |h_1'||x_1(t-1) - x_1(t-2)|,$$

where h_1' is evaluated at an intermediate value between $x_1(t-1)$ and $x_1(t-2)$, while $|h_1'| < 1$ by virtue of Assumptions 5.1.1 and 5.1.2. Since the mapping (7) is a contraction, it has a unique fixed point x_1^* which is globally stable, where x_1^* is the equilibrium output of the leader. (See Mathematical Appendix 1.3.1 for contraction mapping principle).   Q.E.D.

Before closing this section mention should be made of related mathematical works by J.B.Cruz,Jr., C.I.Chen and M.Simaan. They made static and dynamic extensions of the Stackelberg idea of leader-follower in duopoly models to general nonzero sum games as follows: (See Simaan and Cruz(1973) and Chen and Cruz(1972).)

Let U_1 and U_2 be two sets of admissible strategies for player 1 and 2, respectively. Let the i-th player minimize its cost function

$$J_i(u_i, u_j), \quad i \neq j, \ j=1,2.$$

Call the player that selects his strategy first the leader and the player that selects his second the follower. Without loss of generality, let a Stackelberg strategy refer to a Stackelberg strategy with player 2 as the leader.

Definition 5.1.1:   If there exists a mapping T: $U_2 \rightarrow U_1$ such that for any fixed $u_2 \in U_1$,

$$J_1(Tu_2, u_2) \leq J_1(u_1, u_2) \text{ for any } u_1 \in U_1,$$

and if there exists a strategy $u_{2s2} \in U_2$ for which

$$J_2(Tu_{2s2}, u_{2s2}) \leq J_2(Tu_2, u_2) \text{ for any } u_2 \in U_2,$$

then the pair of two players' strategies $(u_{1s2}, u_{2s2}) \in U_1 \times U_2$,

where $u_{1s2} = Tu_{2s2}$, is called a Stackelberg strategy.

<u>Theorem 5.1.2</u>(Simaan and Cruz): If $U_1 \subset R^m$ and $U_2 \subset R^n$ are compact, and
if J_1 and J_2 are real valued continuous functions on $U_1 \times U_2$,
then a Stackelberg strategy exists.

The above theorm states only existence conditions, while Theorem
5.1.1 is concerned with both existence and stability of a Stackelberg
strategy. Assume $p = f(x_1 + x_2) = 0$ for $x_1 + x_2 = Q \geq \bar{Q}$. Then x_1 and
x_2 both belong to compact sets. Since $\pi_i(x_i, x_j)$, the profit function of
the i-th firm, is real valued and continuous by differentiability
assumption, cost functions in Simaan and Cruz's sense are given by
$J_1(x_1, x_2) = -\pi_1(x_1, x_2)$ and $J_2(x_1, x_2) = -\pi_2(x_1, x_2)$. Applying
Theorem 5.1.2, existence of a Stackelberg strategy in our duopoly model
is established.

5.2.   Resolution of Stackelberg Disequilibrium in a Leader-Leader Model[1]

5.2.1.   Introductory Remarks

As was already mentioned, a Stackelberg disequilibrium will be
encountered and the final result will be unpredictable when two firms
in duopoly strive for leadership at the same time. Of course, there
may be a state where both firms acting as leaders choose constant outputs over time as is shown in Okuguchi(1971). Such a state, however,
is not an equilibrium as each firm's expectations on the other firm's
output will continuously turn out to be mistaken. It is, therefore,
rather strange to consider that both firms will continue to behave as
leaders in such a state.

As for a leader-leader model, Stackelberg assumes perfect information in the sense that each firm knows correctly the other's reaction
function, which is tantamount to assuming that each firm has perfect
information of the other's cost and market demand functions. If we
drop this assumption of perfect information for a leader-leader model
of duopoly, letting each duopolist try to search for estimates of
conjectural variations in the sense of Frish[2] things will become much

---

1)   This section is based on T.Negishi and K.Okuguchi(1972).
2)   See Frish(1933). For empirical testing based on data of the
Japanese flat glass industry, see G.Iwata(1974).

better, and the difficulty of indeterminacy of Stackelberg disequilibrium will be removed. Thus as was shown by R.Sato and K.Nagatani(1967)
equilibrium will be reached where both firms' expectations are justified
by what actually happens. The trouble, however, is that everything
will depend on coefficients of subjective conjectural variations. As
was stressed by A.Heertje,[1] these coefficients should have objective
bases, which we hope to give in terms of parameters of cost and market
demand functions.

In this section we depart from Stackelberg's overly mechanical
duopoly model and analyze the (existence and) stability of equilibrium
in a modified leader-leader model where information is not perfect
and each firm estimates successively the other's cost function from
the past knowledge of the two firms' outputs.

5.2.2. Local Stability

In this subsection we shall deal with the existence and local
stability of equilibrium of a modified Stackelberg leader-leader model
to be described below. Let the market demand function be

(1) $p = a - b(x_1 + x_2),$ $a, b > 0,$

where p, x_1 and x_2 have the same meaning as before, and each duopolist
is assumed to have correct knowledge of the market demand function.
If firm 1 acts as the leader, it has to estimate firm 2's reaction
function. Assuming that the marginal cost of firm 2 perceived by firm
1, MC_2^p, be linear

(2) $MC_2^p = k_2 + m_2 x_2,$

where k_2 and m_2 are constants, and letting firm 1 think that firm 2's
marginal revenue and marginal cost are equal for a given value of $x_1(t)$,
the following equation is derived.

(3) $a - b(x_1(t) + x_2) - bx_2 = k_2 + m_2 x_2,$

where $x_1(t)$ is firm 1's output in period t. This is the reaction function of firm 2 as perceived by firm 1. The perceived reaction function
must be consistent with what actually happened in period t-1, thus,

(4) $a - b(x_1(t-1) + x_2(t-1)) - bx_2(t-1) = k_2 + m_2 x_2(t-1).$

From (3) and (4) an explicit form for firm 2's reaction function as
perceived by firm 1 in period t is shown to be

---

1) Heertje(1960). Heertje's solution based on a theory of profit share,
however, differs from ours which is based on Stackelberg's idea.

(5) $\qquad x_2 = x_2(t-1) - (x_1(t) - x_1(t-1))b/(m_2 + 2b),$

which shows that the output of firm 2 in period t perceived by firm 1 is a function not only of firm 1's output in the same period but also of both firms' outputs in period t-1.[1] The first order condition for profit maximization of firm 1 as the leader is given by

(6) $\qquad a - b\{x_1(t) + x_2(t-1) - (x_1(t) - x_1(t-1))b/(m_2 + 2b)\}$

$\qquad - x_1(t)b(m_2 + b)/(m_2 + 2b) - MC_1(x_1(t)) = 0,$

where MC_1 is actual (as distinct from perceived) marginal cost of firm 1 as a function of its output. Equation (6) is a consequence of equating marginal revenue and marginal cost of firm 1 as the leader. Similarly, profit maximizing output of firm 2 in period t as the leader, $x_2(t)$, is derivable from

(7) $\qquad a - b\{x_2(t) + x_1(t-1) - (x_2(t) - x_2(t-1))b/(m_1 + 2b)\}$

$\qquad - x_2(t)b(m_1 + b)/(m_1 + 2b) - MC_2(x_2(t)) = 0,$

where m_1 corresponds to m_2 of firm 1 and MC_2 is the actual marginal cost of firm 2. The dynamics of duopoly when both firms strive for leadership is represented by a system of difference equations (6) and (7).

Let us consider a stationary state of (6) and (7) whose outputs are x_1^* and x_2^*. For these outputs, conjectures of both firms acting as leaders are justified by virtue of (5) and the corresponding expression for firm 2, and no conjecture revisions are made. Of course, even in such a stationary state, conjectures can only be quasi-correct in the sense of Fellner. They are right for the wrong reason and are justified on the bases of entirely arbitrary notions of what a firm would do if the other firm changed its output. Fellner(1965) insists that the stationary state can be continued only as long as nobody realizes that their notions are incorrect, and that it is extremely unlikely that no one should ever test them by moving out of the stationary state to see that the other firm would not react along its alleged reaction function. However, if a firm does deviate from its behavior as the leader to conduct such a test, it must expect that the other firm may also deviate temporarily from its reaction function to see whether its behavior based on the supposition of unchanged rival's output is

---

1) For a similar device in a general equilibrium model under monopolistic competition known as a perceived demand function, see Negishi (1961).

justified. Such a test, therefore, can not easily convince the leader that its basic assumption that the rival is a follower is not valid and it is likely that the stationary solution is viable.

When both MC_1 and MC_2 are linear, existence of positive equilibrium outputs x_i^*'s is easily proved under reasonable assumptions on relevant parameters. In order to derive a necessary and sufficient condition for local stability of the stationary state or equilibrium in a modified Stackelberg leader-leader duopoly under no product differentiation, expand (6) and (7) at equilibrium to get,

$$y_1(t) = A_{11}^* y_1(t-1) + A_{12}^* y_2(t-1),$$

(8)

$$y_2(t) = A_{21}^* y_1(t-1) + A_{22}^* y_2(t-1),$$

where $y_i(t) \equiv x_i(t) - x_i^*$, $\qquad i=1,2$,

$$A_{11}^* \equiv -b^2/(2b^2 + 2bm_2 + \alpha_1^* + 2b\alpha_1^*),$$

$$A_{12}^* \equiv -b(m_2 + 2b)/(2b^2 + 2bm_2 + \alpha_1^* m_2 + 2b\alpha_1^*),$$

$$A_{21}^* \equiv -b(m_1 + 2b)/(2b^2 + 2bm_1 + \alpha_2^* m_1 + 2b\alpha_2^*),$$

$$A_{22}^* \equiv -b^2/(2b^2 + 2bm_1 + \alpha_2^* m_1 + 2b\alpha_2^*),$$

$$\alpha_i^* \equiv MC_i'^* = dMC_i/dx_i \big|_{x_i = x_i^*}, \qquad i=1,2.$$

We assume that m_i's and $MC_i'^*$'s are both non-negative, thus, $A_{ij}^* < 0$ $i,j=1,2$. A characteristic equation for (8) given by

(10) $\qquad \lambda^2 - (A_{11}^* + A_{22}^*)\lambda + A_{11}^* A_{22}^* - A_{12}^* A_{21}^* = 0$

has two real roots as is easily seen. Since

$$A_{11}^* + A_{22}^* < 0, \ |A_{11}^*| < |A_{12}^*|, \ |A_{22}^*| < |A_{21}^*|,$$

one of these roots is negative and has a larger absolute value than the other positive root. Thus, stability of (8) follows provided the absolute value of the negative root is less than unity.

Theorem 5.2.2.1: The equilibrium for (8) is stable if and only if

(11) $\quad C(m_1, m_2, \alpha_1^*, \alpha_2^*) = 1 + (A_{11}^* + A_{22}^*) + A_{11}^* A_{22}^* - A_{12}^* A_{21}^* > 0$

Under the assumption of $m_i \geq 0$ and $\alpha_i^* \geq 0$, we can easily show that

$$\partial C/\partial \alpha_i^* > 0, \ \partial C/\partial m_i > 0, \qquad i=1,2.$$

If $m_1 = m_2 = \alpha_1^* = \alpha_2^* = b$, we have

$$A_{11}^* = A_{22}^* = -/7, \ A_{12}^* = A_{21}^* = -3/7,$$

yielding $C > 0$. Thus if m_1's and α_i^*'s are all positive and large, relative to b, stability of (8) follows. Simce m_1 (or m_2) is an estimate of α_1^* (or α_2^*), we may reasonably suppose that α_1^* (or α_2^*) is positive if marginal cost of firm 1 (or firm 2) is actually increasing. However, $C < 0$ and instability follows provided $m_1 = m_2 = \alpha_1^* = \alpha_2^* = 0$. Summarizing,

Proposition 5.2.2.1: The equilibrium for a system of difference equations (6) and (7) is locally stable if marginal costs of both firms are increasing sharply, relative to the slope of the market demand curve.

5.2.3. Global Stability

Global stability can be analyzed as follows. Assume that $x_i's$ take non-negative and unique values against any $x(t-1) = (x_1(t-1), x_2(t-1))$. Thus from (6) and (7),

(12) $x_1(t) = g^1(x_1(t-1), x_2(t-1))$, $i=1,2$,

where g^1's are assumed to be differentiable. From (6) and (7) we also have,

(13) $\partial x_1(t)/\partial x_j(t-1) \equiv A_{1j}$, $i,j=1,2$,

where A_{1j}'s equal corresponding ones without asterisks in (9), with α_1 evaluated at $x_1(t-1)$ $(i=1,2)$, and are negative provided m_1's and α_1's are non-negative. Suppress t-1 in (12), and introduce

Assumption 5.2.3.1: For any x_1 and x_2,

(14) $|\partial g^1/\partial x_1| + |\partial g^2/\partial x_1| < 1$,

(15) $|\partial g^1/\partial x_2| + |\partial g^2/\partial x_2| < 1$,

hold simultaneously.

We then have

Theorem 5.2.3.1: Under Assumption 5.2.3.1 equilibrium exists and is globally stable.

Proof: The proof which we omit is completed by showing that (12) is a contraction.   Q.E.D.

Stability conditions (14) and (15) are equivalent to (16) and (17), respectively.

(16) $D^1(m_1, m_2, \alpha_1, \alpha_2) \equiv \{2b(m_2 + b) + \alpha_1(m_2 + 2b)\}\{2b(m_1 + b)$

$+ \alpha_2(m_1 + 2b)\} - b^2\{2b(m_1 + b) + \alpha_2(m_1 + 2b)\} - b(m_1 + 2b)$

$\{2b(m_2 + b) + \alpha_1(m_2 + 2b)\} > 0$,

(17) $D^2(m_1, m_2, \alpha_1, \alpha_2) \equiv \{2b(m_2 + b) + \alpha_1(m_2 + 2b)\}\{2b(m_1 + b)$

$+ \alpha_2(m_1 + 2b)\} - b^2\{2b(m_2 + b) + \alpha_1(m_2 + 2b)\} - b(m_2 + 2b)$

$\{2b(m_1 + b) + \alpha_2(m_1 + 2b)\} > 0.$

Consider a special case where $m_1 = m_2 = \alpha_1 = \alpha_2 = b$. Then $D^1 = D^2 = 21b^4 > 0$, implying global stability. To consider a more general situation, note

(18) $\partial D^1/\partial m_j > 0, \ \partial D^1/\partial \alpha_j > 0, \qquad i,j = 1,2.$

Taking into account the result for the special case above and (18), we can assert

Proposition 5.2.3.1: The equilibrium for (12) is globally stable pro-
                      vided perceived as well as actual marginal costs of the two
                      firms are sharply increasing, relative to the slope of the
                      market demand curve.

5.2.4. Unknown Market Demand Function and Global Stability

    In our analyses so far, market demand function was perfectly known to both firms in duopoly. We shall now extend our analyses to a case where each firm is assumed to subjectively perceive market demand function. The perceived demand function to be shortly introduced was first considered by Y.Hosomatsu(1969) in the context of a Cournot oligopoly model under no product differentiation. Let

(19) $p = a_i - b_i(x_1 + x_2), \ a_i, \ b_i > 0, \qquad i = 1,2,$

be the i-th firm's perceived market demand function, where a_i's and b_i's remain constant over time. Following a similar line of reasoning as was employed in deriving (6) and (7) and rearranging the resulting expressions,

(20) $2b_1(m_2 + b_1)/(m_2 + 2b_1) \cdot x_1(t) + MC_1(x_1(t)) =$

        $a_1 - b_1^2/(m_2 + 2b_1) \cdot x_1(t-1) - b_1 x_2(t-1),$

(21) $2b_2(m_1 + b_2)/(m_1 + 2b_2) \cdot x_2(t) + MC_2(x_2(t)) =$

        $a_2 - b_2 x_1(t-1) - b_2^2/(m_1 + 2b_2) \cdot x_2(t-1).$

Suppose as before that $x_1(t)$'s are non-negative, single-valued and differentiable. Hence,

(22) $x_1(t) = h^1(x_1(t-1), x_2(t-1)), \qquad i = 1,2,$

where

(23) $\partial x_i/\partial x_i(t-1) = -b_i^2/\{2b_i(m_j + b_i) + \alpha_i(m_j + 2b_i)\}, i \neq j, i,j=1,2.$

(24) $\partial x_i(t)/\partial x_j(t-1) = -b_i(m_j + 2b_i)/\{2b_i(m_j + b_i) + \alpha_i(m_j + 2b_i)\},$

$$i \neq j, \quad i,j=1,2.$$

Expressions (23) and (24) become negative provided m_i's and α_i's are non-negative. We can prove by showing that (22) is a contraction, the following

Theorem 5.2.4.1:   The equilibrium for (22) exists and is globally stable provided

(25) $|\partial h^1/\partial x_j| + |\partial h^2/\partial x_j| < 1, \qquad j=1,2.$

Taking into account (23) and (24), (25) is shown to be true if

(26) $E^1(m_1, m_2, b_1, b_2, \alpha_1, \alpha_2) \equiv \{2b_1(m_2 + b_1) + \alpha_1(m_2 + 2b_1)\}$

$\{2b_2(m_1 + b_2) + \alpha_2(m_1 + 2b_2)\} - b_1^2\{2b_2(m_1 + b_2) + \alpha_2(m_1 + 2b_2)\}$

$- b_2(m_1 + 2b_2)\{2b_1(m_2 + b_1) + \alpha_1(m_2 + 2b_1)\} > 0,$

(27) $E^2(m_1, m_2, b_1, b_2, \alpha_1, \alpha_2) > 0,$

are satisfied simultaneously, where $E^2(\cdot)$ is obtained from (26) by replacing m_1, α_2, b_1 and b_2 in the second and third terms in (26) with m_2, α_1, b_2 and b_1, respectively.
    Consider now a simple case where

(28) $m_i = \alpha_i = b_i, \qquad i=1,2.$

After a little manipulation,

$$E^1 = b_i b_j (5b_i^2 + 10b_i b_j + 6b_j^2) > 0, \qquad i \neq j, \ i,j=1,2,$$

which leads to global stability. By calculating further,

$$\partial E^1/\partial m_j > 0, \ \partial E^1/\partial \alpha_j > 0, \qquad i,j=1,2,$$

which in combination with global stability established for the case of (28) enables us to claim

Proposition 5.2.4.1:   The equilibrium for (22) is globally stable if marginal cost of firm 1(2) perceived by firm 2(1) and actual marginal cost of firm 1(2) are increasing sharply in comparison with the slope of market demand curve perceived by firm 1(2).

CHAPTER 6

EXTRAPOLATIVE EXPECTATIONS AND STABILITY OF OLIGOPOLY EQUILIBRIUM

6.1. Introduction

The Cournot assumption on rivals' behavior in oligopoly is un-
doubtedly naive. Two more general formulae for formation of expectations
exist in economic theory. One is extrapolative and the other adaptive.
Under the former, the expected value of an economic variable x in period
t, x_t^e, is given by

$$x_t^e = x_{t-1} + \alpha(x_{t-1} - x_{t-2}), \quad |\alpha| < 1 \quad \text{(discrete version)},$$

or

$$x_t^e = x_t + \alpha dx_t/dt \quad \text{(continuous version)}.$$

A.C.Enthoven and K.J.Arrow(1956) assume non-negativity of α, but such
an assumption is not necessary. If $\alpha = 0$, extrapolative expectations
turn out to be the same as the Cournot assumption in both discrete and
continuous versions. In adaptive expectations, it is assumed that they
are formed according to

$$x_t^e = x_{t-1}^e + \beta(x_{t-1} - x_{t-1}^e), \quad 0 < \beta \leq 1 \quad \text{(discrete version)}$$

or

$$dx_t^e/dt = \beta(x_t - x_t^e), \quad \beta > 0 \quad \text{(continuous version)}.$$

If $\beta = 1$ (discrete version) or $\beta = \infty$ (continuous version), the Cournot
assumption emerges. Adaptive expectations were first introduced by
P.Cagan(1956) to explain post-war German hyper-inflation, and then used
by M.Nerlove(1958) to analyze American agriculture between 1909 and 1932.
According to adaptive expectations, expected values are successively
revised in the same direction as the difference between actual and expected values. From the discrete version the following expression
results if t is sufficiently large.

$$x_t^e = \beta \sum_{i=1}^{\infty} (1 - \beta)^{i-1} x_{t-1}.$$

Thus the expected value is a weighted average of all actual values in
the past, the weight declining as time recedes. This point can also
been seen by integrating the expression for the continuous version.
Rewriting the discrete version as

$$x_t^e = \beta x_{t-1} + (1 - \beta) x_{t-1}^e,$$

it is seen that the expected value in period t is a weighted average of
actual and expected values in period t-1.

The implications of these two formulae of expectations for stability
of competitive equilibrium in a Walrasian general equilibrium model have
been investigated by Enthoven and Arrow(1956), Arrow and Nerlove(1958),
T.Negishi(1964) and Arrow and Hurwicz(1962).

We owe to R.E.Quandt(1967) for the first systematic treatment of
stability of oligopoly equilibrium in relation to extrapolative expectations in a model of no product differentiation. His result was extended
by Okuguchi(1969) by taking into consideration the effects of alternative
expectations, that is, adaptive expectations on stability of oligopoly
equilibrium also for the case of no product differentiation.

In this chapter effects of extrapolative expectations on stability
of oligopoly equilibrium are explored in detail both when product differentiation exists and when it does not. Following Quandt(1967), we
shall be interested in whether a sufficient condition for stability of
the Cournot oligopoly solution will entail stability of oligopoly equilibrium in models characterized by introduction of extrapolative expectations of firms regarding rivals' prices or outputs.

6.2. Stability under No Product Differentiation

In this section it is assumed that no product differentiation exists.
Let there be n firms, and for the sake of simplicity, assume that the
i-th and j-th firms' expectations on the k-th firm's output ($i \neq j \neq k$) are
the same. Since all firms are assumed to form expectations on rivals'
outputs extrapolatively,

(1) $\qquad x_i^e = x_i + \lambda_i dx_i/dt, \qquad i=1,2,\cdots,n,$

where x_i and x_i^e are actual and expected (by rivals) outputs of the i-th
firm. The market demand function, about which all firms have complete
knowledge, is assumed to be linear:

(2) $\qquad p = a - b\Sigma_i x_i, \qquad a, b > 0,$

The i-th firm's cost function is quadratic,

(3) $\qquad C_i(x_i) = e_i + c_i x_i + d_i x_i^2/2, \qquad i=1,2,\cdots,n.$

Since the i-th firm's profit function is defined by

(4) $\qquad \pi_i(x_i, x_{-i}) \equiv px_i - C_i(x_i), \qquad i=1,2,\cdots,n,$

maximization of the i-th firm's expected profit will yield,

(5) $\qquad x_i^* = \eta_i + \xi_i \sum_{j \neq i} x_j^e, \qquad i=1,2,\cdots,n,$

where x_i^* is the i-th firm's expected profit maximizing output, and

$$\eta_i \equiv (a - c_i)/(d_i + 2b), \quad \xi_i \equiv -b/(d_i + 2b), \quad i=1,2,\cdots,n.$$

From the second order condition for expected profit maximization as given by

(6) $\qquad d_i + 2b > 0, \qquad i=1,2,\cdots,n,$

we have $\xi_i < 0$, while $\eta_i > 0$ as x_i^* must be positive when $x_{-i}^e = 0$. Actual output is not necessarily adjusted instantaneously to expected profit maximizing output, thus

(7) $\qquad dx_i/dt = k_i(x_i^* - x_i), \quad k_i > 0, \qquad i=1,2,\cdots,n.$

From (1), (5) and (7)

(8)

$$
\begin{bmatrix} dx_1/dt \\ dx_2/dt \\ \vdots \\ dx_n/dt \end{bmatrix}
=
\begin{bmatrix}
-k_1 & k_1\xi_1 & \cdot & k_1\xi_1 \\
k_2\xi_2 & -k_2 & \cdots & k_2\xi_2 \\
\vdots & \vdots & \ddots & \vdots \\
k_n\xi_n & k_n\xi_n & & -k_n
\end{bmatrix}
\begin{bmatrix} x_1 \\ x_2 \\ \vdots \\ x_n \end{bmatrix}
$$

$$
+
\begin{bmatrix}
0 & k_1\xi_1 & \cdots & k_1\xi_1 \\
k_2\xi_2 & 0 & \cdots & k_2\xi_2 \\
\vdots & & \ddots & \vdots \\
k_n\xi_n & k_n\xi_n & \cdots & 0
\end{bmatrix}
\begin{bmatrix} \lambda_1 dx_1/dt \\ \lambda_2 dx_2/dt \\ \vdots \\ \lambda_n dx_n/dt \end{bmatrix}
+
\begin{bmatrix} k_1\eta_1 \\ k_2\eta_2 \\ \vdots \\ k_n\eta_n \end{bmatrix}
$$

To let (8) be more compact by using matrix notation, define:

$$
K = \begin{bmatrix} k_1 & & & \\ & k_2 & & \\ & & \ddots & \\ & & & k_n \end{bmatrix}, \quad
A = \begin{bmatrix} 0 & \xi_1 & \cdot & \xi_1 \\ \xi_2 & 0 & \cdot & \xi_2 \\ \vdots & \vdots & \ddots & \vdots \\ \xi_n & \xi_n & \cdots & 0 \end{bmatrix},
$$

$$
\Lambda = \begin{bmatrix} \lambda_1 & & & \\ & \lambda_2 & & \\ & & \ddots & \\ & & & \lambda_n \end{bmatrix},
$$

$$I = \begin{bmatrix} 1 & & & & \\ & 1 & & 0 & \\ & & 1 & & \\ & 0 & & \ddots & \\ & & & & \ddots \\ & & & & & 1 \end{bmatrix}$$

$$\gamma = (k_1\eta_1, k_2\eta_2, \cdots, k_n\eta_n)'$$

$$x = (x_1, x_2, \cdots, x_n)'.$$

$$dx/dt = (dx_1/dt, dx_2/dt, \cdots, dx_n/dt)'.$$

We can rewrite (8) as

(9) $dx/dt = (I - K A \Lambda)^{-1} K(A - I)x + (I - K A \Lambda)^{-1}\gamma.$

If $\Lambda = 0$, (9) reduces to none other than the Cournot model,

(10) $dx/dt = k(A - I)x + \gamma.$

Noting that diagonal elements of $K(A - I)$ are all negative, and applying McKenzie's theorem,[1] the stability of the Cournot oligopoly solution follows provided,

(11) $\xi_i > -1/(n - 1)$, $i = 1, 2, \cdots, n.$

The problem we are interested in is whether (11) implies stability of
the oligopoly solution for the dynamic system (9). To analyze this,
assume $K = I$,

(12) $0 \le \lambda_i < 1$, $i = 1, 2, \cdots, n$,[2]

(13) $D = [d_{ij}] = (I - A\Lambda)^{-1} \geqq 0.$

Taking into account an identity

(14) $F \equiv (I - A\Lambda)^{-1}(A - I) = -I + (I - A\Lambda)^{-1}A(I - \Lambda),$

let

$$B = [b_{ij}] = (I - A\Lambda)^{-1}A(I - \Lambda)$$

$$C = [c_{ij}] = -B.$$

Since B is non-negative, it has the Frobenius root $\rho(C)$. We have to
analyze two cases depending on whether all λ_i's are identical or not.

---

1) See L.W. McKenzie (1960, Theorem 2, p. 49) which runs: If a square
matrix has a quasi-dominant negative diagonal, real parts of its characteristic roots are all negative.
2) If $-1 < \lambda_i \le 0$ for all i, $A\Lambda$ becomes non-negative, and by virtue of
(11), its Frobenius root is less than one and $(I - A\Lambda)^{-1} \geqq 0.$

Assume first $\lambda_i = \lambda$ for all i. Taking into consideration (11) it follows,

(15) $\qquad \sum_j c_{ij} = -(1 - \lambda)(n - 1)\sum_j d_{ij}$

$\qquad\qquad\qquad < (1 - \lambda)\sum_j d_{ij}, \qquad i=1,2,\cdots,n.$

For any matrix $G = [g_{ij}]$, define (maximum row sum) norm $\|G\|$ by

$$\|G\| = \max_i \sum_j |g_{ij}|.$$

From (11) $\|A\| < 1$, therefore,

$$\|D\| = \max_i \sum_j |\alpha_{ij}| = \| I + \lambda A + \lambda^2 A^2 + \lambda^3 A^3 + \cdots \|$$

(16) $\qquad\qquad\qquad\qquad \leq 1 + |\lambda|\|A\| + |\lambda^2|\|A\|^2 + |\lambda^3|\|A\|^3 + \cdots$

$\qquad\qquad\qquad\qquad \leq 1 + \lambda + \lambda^2 + \lambda^3 + \cdots$

$\qquad\qquad\qquad\qquad = 1/(1 - \lambda).$

Taking into account $\sum_j c_{ij} < 1$ for all i which is a consequence of (15) and (16),

(17) $\qquad \rho(C) < 1$

is derived. Thus B has a negative characteristic root whose largest modulus is less than one. This fact coupled with (14) enables one to conclude that real parts of all characteristic roots of F are negative, proving stability.

We next shall consider a more general case where at least two λ_i's take different values. Let, without loss of generality,

$$\Lambda = \begin{bmatrix} \lambda_1^* & & & & & & & \\ & \ddots \lambda_1^* & & & & 0 & & \\ & & \lambda_2^* & & & & & \\ & & & \ddots \lambda_2^* & & & & \\ & & & & & \ddots & & \\ & 0 & & & & \lambda_s^* & & \\ & & & & & & \lambda_s^* & \end{bmatrix} \begin{matrix} \left.\vphantom{\begin{matrix}a\\a\end{matrix}}\right\} n_1 \\[2ex] \left.\vphantom{\begin{matrix}a\\a\end{matrix}}\right\} n_2 \\[2ex] \left.\vphantom{\begin{matrix}a\\a\end{matrix}}\right\} n_s \end{matrix}$$

$$n_1 + n_2 + \cdots + n_s = n,$$

$$\min_i \lambda_i = \lambda_1^* < \lambda_2^* < \cdots < \lambda_s^* = \max_i \lambda_i$$

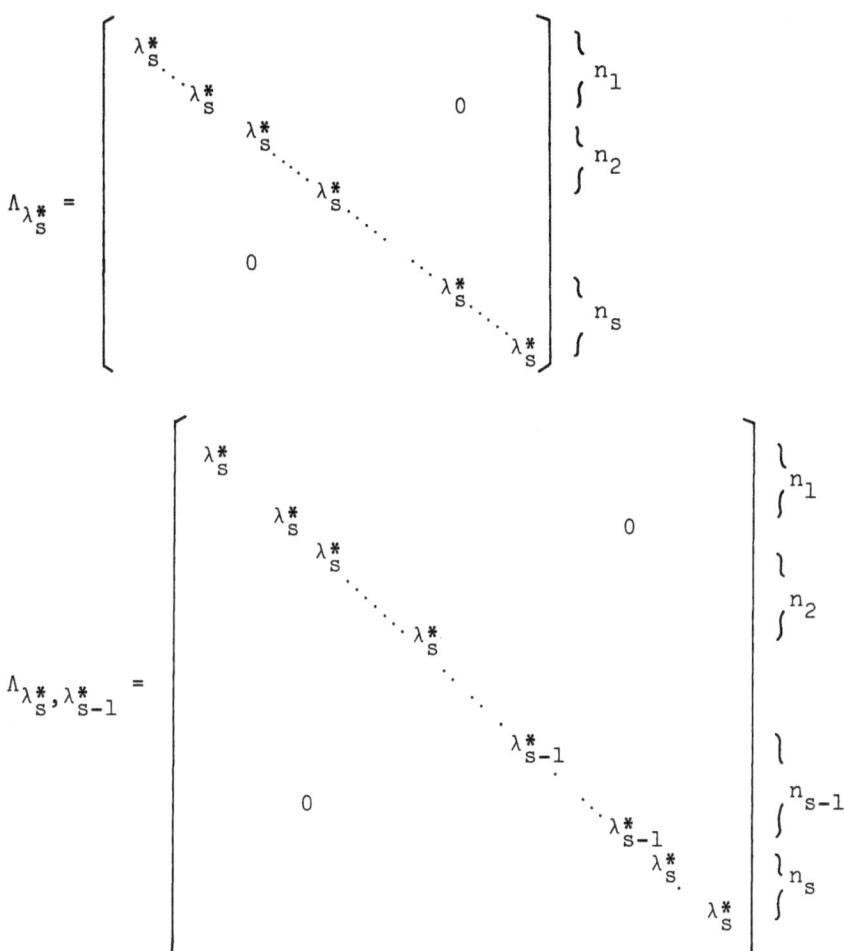

etc.. Thus by definition,

$$\Lambda_{\lambda_s^*, \lambda_{s-1}^*, \ldots, \lambda_1^*} = \Lambda.$$

From the result proved for the case of identical λ_i's, we see

$$\rho(-(I - A\Lambda)^{-1}A(I - \Lambda_{\lambda_s^*})\,) < 1.$$

By a property of the Frobenius root that it is non-decreasing in any element of a non-negative square matrix,

$$\rho(-(I - a\Lambda)^{-1}A(I - \Lambda_{\lambda_s^*, \lambda_{s-1}^*})\,)\,) \leq \rho(-(I - A\Lambda)^{-1}\,A(I - \Lambda_{\lambda_s^*})\, < 1.$$

Arguing similarly, we finally arrive at

(15) $\rho(-(I - A\Lambda)^{-1}A(I - \Lambda)\,) < 1,$

which proves that real parts of all characteristic roots of F are negative, hence stability.

The case where each firm has extrapolative expectations on the rest of industry output can be analyzed as follows. Let $x^i = \sum_{j \neq i} x_j$ be the rest of industry output for the i-th firm, and x^{ie} its expectation. The dynamic system now consists of (16) - (18) below.

(16) $dx_i/dt = k_i(x_i^* - x_i),$ $i=1,2,\cdots,n,$

(17) $x^{ie} = x^i + \mu_i dx^i/dt,$ $i=1,2,\cdots,n,$

(18) $x_i^* = n_i + \xi_i x^{ie},$ $i=1,2,\cdots,n.$

In matrix notation,

$$
\begin{bmatrix}
1 & -k_1\xi_1\mu_1 & \cdots\cdots & -k_1\xi_1\mu_1 \\
-k_2\xi_2\mu_2 & 1 & \cdots\cdots & -k_2\xi_2\mu_2 \\
\vdots & & \ddots & \vdots \\
-k_n\xi_n\mu_n & -k_n\xi_n\mu_n & \cdots\cdots & 1
\end{bmatrix}
\begin{bmatrix}
dx_1/dt \\
dx_2/dt \\
\vdots \\
dx_n/dt
\end{bmatrix}
=
$$

(19)

$$
\begin{bmatrix}
-k_1 & k_1\xi_1 & \cdots\cdots & k_1\xi_1 \\
k_2\xi_2 & -k_2 & \cdots\cdots & k_2\xi_2 \\
\vdots & & \ddots & \vdots \\
k_n\xi_n & k_n\xi_n & \cdots\cdots & -k_n
\end{bmatrix}
\begin{bmatrix}
x_1 \\
x_2 \\
\vdots \\
x_n
\end{bmatrix}
+
\begin{bmatrix}
k_1 n_1 \\
k_2 n_2 \\
\vdots \\
k_n n_n
\end{bmatrix}
$$

Define A, γ, x and dx/dt as before and let

$$
M =
\begin{bmatrix}
\mu_1 & & & & 0 \\
& \mu_2 & & & \\
& & \ddots & & \\
0 & & & \ddots & \\
& & & & \mu_n
\end{bmatrix}
$$

If K = I, (19) is simplified as

(20) $dx/dt = (I - MA)^{-1}(A - I)X + (I - MA)^{-1}Y.$

Assume $0 \leq \mu_i < 1$ for all i and non-negativity of $(I - MA)^{-1}$. Stability of the oligopoly solution for (20) may be established based on similar arguments employed for proving stability of (9), and taking into account an identity

$$(I - MA)^{-1}(A - I) = -I + (I - MA)^{-1}(I - M)A.$$

6.3. Product Differentiation and Stability

In this section extrapolative expectations are introduced into a model with product differentiation. Let p_i^e be expected price for the i-th firm's product. Demand function for the i-th firm and its cost function are

(1) $x_i = a_i - \sum_j b_{ij} p_j,$ i=1,2,$\cdots$,n,

(2) $C_i(x_i) = e_i + c_i x_i + d_i x_i^2/2,$ i=1,2,$\cdots$,n,

where

$a_i > 0, \; b_{ii} > 0, \; b_{ij} < 0$ $i \neq j, \; i,j=1,2,\cdots,n.$

$c_i \geq 0, \; d_i \geq 0, \; e_i \geq 0$

Profit function of the i-th firm is

(4) $\pi_i(p) = \pi_i(p_i, p_{-i}) = p_i x_i - C_i(x_i),$ i=1,2,$\cdots$,n.

Let p_i^* be the i-th firm's expected price maximizing price. We then have,

(5) $p_i^* = \alpha - \sum_{j \neq i} r_{ij} p_j^e,$ i=1,2,$\cdots$,n,

$\alpha_i \equiv (a_i + b_{ii} c_i + a_i b_{ii} d_i)/(2 + b_{ii} d_i) b_{ii},$ i=1,2,$\cdots$,n,

$r_{ij} \equiv (1 + b_{ii} d_i) b_{ij}/(2 + b_{ii} d_i) b_{ii},$ $i \neq j, \; i,j=1,2,\cdots,n.$

The second order condition together with (3) ensures $\alpha_i < 0$ and $r_{ij} < 0$ for all i and j, $i \neq j$. Since expectations are extrapolative,

(6) $p_i^e = p_i + \lambda_i dp_i/dt,$ i=1,2,$\cdots$,n.

Actual price is assumed to be adjusted with a lag according to

(7) $dp_i/dt = k_i(p_i^* - p_i), \; k_i > 0,$ i=1,2,$\cdots$,n.

(5), (6) and (7) together yield

(8) $\quad dp_i/dt = -(I + KR\Lambda)^{-1}K(I + R)p + (I + KR\Lambda)K\alpha,$

$$R = \begin{bmatrix} 0 & r_{12} \cdots \cdots \cdots r_{1n} \\ r_{21} & 0 \cdots \cdots \cdots r_{2n} \\ \vdots & \vdots \\ \vdots & \vdots \\ r_{n1} & r_{n2} \cdots \cdots 0 \end{bmatrix},$$

$$K = \begin{bmatrix} k_1 & & & \\ & k_2 & & 0 \\ & & \ddots & \\ 0 & & & \\ & & & k_n \end{bmatrix}, \quad \Lambda = \begin{bmatrix} \lambda_1 & & & \\ & \lambda_2 & & 0 \\ & & \ddots & \\ 0 & & & \\ & & & \lambda_n \end{bmatrix}$$

$p = (p_1, p_2, \cdots, p_n)',$

$\alpha = (\alpha_1, \alpha_2, \cdots, \alpha_n)'.$

Under the Cournot assumption, $\Lambda = 0,$

(9) $\quad dp/dt = -K(I + R)p + K\alpha.$

Taking into account that off-diagonal elements of $-K(I + R)$ are all positive, and applying McKenzie's theorem on a matrix with quasi-dominant diagonals[1], the price adjusting oligopoly solution for (9) is proved to be stable _if_ _and_ _only_ _if_

(10) $\quad \sum\limits_{j \neq i} r_{ij} > -1, \qquad i=1,2,\cdots,n.$ [2]

Some comments on the stability condition (10) follow.[3] Since $r_{ij} < 0,$ $i \neq j,$ (10) is more likely to be satisfied, the larger r_{ij} is, that is,

---

1) McKenzie(1960, Theorem 2', p.58): Let A be a matrix with all off-diagonal elements non-negative. Then A's characteristic roots have all negative real parts if and only if A has a negative quasi-dominant diagonal.
2) Quandt(1967) assumes $b_{ij} = b_i,$ $i \neq j$ and derives a sufficient condition

$\qquad r_i \equiv (1 + b_{ii}d_i)b_i/(2 + b_{ii}d_i)b_{ii} > -\dfrac{1}{n-1}, \qquad i=1,2,\cdots,n.$
3) See Quandt(1967) and D.Tarr(1975).

the smaller r_{ij} is in absolute value. The following signs on derivatives are immediate:

$$\partial r_{ij}/\partial d_i < 0, \ \partial r_{ij}/\partial b_{ii} > 0, \ \partial r_{ij}/\partial b_{ij} > 0, \quad i \neq j, \ i,j=1,\cdots,n.$$

From these inequalities we know that <u>ceteris paribus</u>, the price adjusting Cournot oligopoly solution in a model of product differentiation is more likely to be stable in any of the following situations.

(a) A smaller rate of change in marginal cost for each firm.

(b) A more sensitive demand for a firm with respect to change in its own price.

(c) A less sensitive demand for a firm to change to other firm's prices.

Inequality (10) is assumed to be true in the following discussion. Now, for the sake of simplicity, let

(11) $K = I$.

From (1) and (11) we have $\rho(-R\Lambda) < 1$, thus

$$G = (I + R\Lambda)^{-1} \geq 0.$$

Note the following rewriting of J and definition of a non-negative matrix H.

$$J \equiv -(I + R\Lambda)^{-1}(I + R) = -I + (I + R\Lambda)^{-1}R(\Lambda - I)$$

$$H = [h_{ij}] = (I + R\Lambda)^{-1}R(\Lambda - I).$$

Due to a property of the Frobenius root,

(12) $\rho(H) \leq \max_i \sum_j h_{ij}.$

Two cases have to be considerd. First let $\lambda_i = \lambda$ for all i. We then have

(13) $\sum_j h_{ij} = (\lambda - 1)\sum_j g_{ij} \sum_{k \neq j} r_{jk}$

$< (1 - \lambda)\sum_j g_{ij}, \quad i=1,2,\cdots,n,$

(14) $\|G\| = \max_i \sum_j g_{ij} < 1/(1 - \lambda).$

From (12) - (14), $\rho(H) < 1$. The same inequality can be derived even when λ_i's have at least two distinct values by arguing similarly as we did in deriving (15) in Section 6.2. In sum, under our assumptions real parts of characteristic roots of J are all negative, which completes our proof of stability of the price adjusting oligopoly solution in a model of product differentiation characterized by extrapolative expectations.

CHAPTER 7

ADAPTIVE EXPECTATIONS AND STABILITY OF OLIGOPOLY EQUILIBRIUM

7.1. No Product Differentiation

In this section we shall analyze stability of equilibrium in an oligopoly model of no product differentiation in which each firm is assumed to have adaptive expectations on all others' outputs. The market demand function, and the cost and profit functions of the i-th firm are given, respectively, by

(1) $p = a - b\Sigma_i x_i, \quad a > 0, \ b > 0,$

(2) $C_i = e_i + c_i x_i + d_i x_i^2/2, \quad c_i \geq 0, \ e_i \geq 0, \qquad i=1,2,\cdots,n$

(3) $\pi_i(x_i, x_{-i}) = px_i - C_i, \qquad i=1,2,\cdots,n.$

No sign restriction is imposed on parameter d_i of the cost function. If d_i is negative, however, marginal as well as total costs might become negative for a sufficiently large output, which is absurd. We, therefore, assume away this absurd situation for profit maximizing output x_i^* for all i.

The i-th firm's expected profit maximizing output under the Cournot assumption is

(4) $x_i^* = \eta_i + \xi_i x^i, \qquad i=1,2,\cdots,n,$

(5)
$$\xi_i \equiv -b/(d_i + 2b),$$
$$\eta_i \equiv (a - c_i)/(d_i + 2b), \qquad i=1,2,\cdots,n.$$

The second order condition

(6) $d_i + 2b > 0, \qquad i=1,2,\cdots,n$

implies $\xi_i < 0$. On the other hand, $\eta_i > 0$ as expected profit maximizing output of the i-th firm has to be positive provided the rivals' outputs are all expected to be zero. Actual output changes with a lag according to

(7) $dx_i/dt = k_i(x_i^* - x_i), \quad k_i > 0, \qquad i=1,2,\cdots,n.$

Adaptive expectations can be introduced as follows. Let x_{ij}^e, $i \neq j$, be the i-th firm's expectation on the j-th firm's output. Movement of x_{ij}^e is governed by

(8) $dx_{ij}^e/dt = m_{ij}(x_i - x_{ij}^e)$, $m_{ij} > 0$, $i \neq j$, $i,j=1,\cdots,n$.

Expected profit maximization for the i-th firm yields

(9) $x_i^* = n_i + \xi_i \sum\limits_{j \neq i} x_{ij}^e$, $i=1,2,\cdots,n$.

Actual output is adjusted according to (7), which coupled with (8) and (9) leads to a system of differential equations (10). The coefficient matrix of (10) can be partitioned as

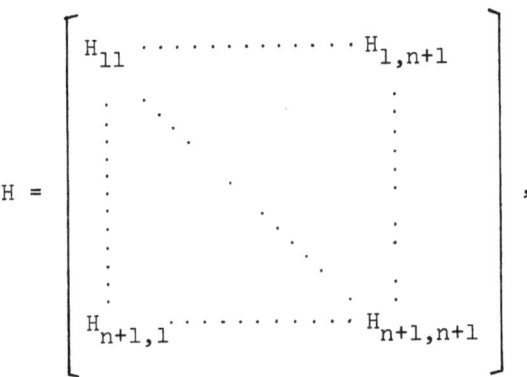

$$H =$$

where

H: $n^2 \times n^2$, H_{11}: $n \times n$,

H_{1j}: $n \times (n-1)$, $j=2,\cdots,n$,

H_{i1}: $(n-1) \times n$, $i=2,\cdots,n$,

H_{ii}: $(n-1) \times (n-1)$ diagonal matrix, $i=2,\cdots,n+1$,

H_{ij}: $(n-1) \times (n-1)$ zero matrix, $i \neq j$, $i,j=2,\cdots,n+1$.

Note that all diagonal elements of H are negative, and assume

(11) $\xi_i > -1/(n-1)$, $i=1,2,\cdots,n$.

Under assumption (11) H has quasi-dominant negative diagonals, implying negativity of real parts of all characteristic roots of H. This proves stability of the solution for (10) under (11).

$$
\begin{bmatrix}
dx_1/dt \\
dx_n/dt \\
dx^e_{12}/dt \\
dx^e_{1n}/dt \\
\vdots \\
dx^e_{n1}/dt \\
\vdots \\
dx^e_{nn-1}/dt
\end{bmatrix}
=
[\,M\,]
\begin{bmatrix}
x_1 \\
x_n \\
x^e_{12} \\
x^e_{1n} \\
\vdots \\
x^e_{n1} \\
\vdots \\
x^e_{nn-1}
\end{bmatrix}
+
\begin{bmatrix}
k_1 n_1 \\
k_n n_n \\
0 \\
0 \\
\vdots \\
0 \\
\vdots \\
0
\end{bmatrix}
$$

where the matrix $[\,M\,]$ contains the elements

$$
-k_1,\quad k_1\xi_1 \cdots k_1\xi_1,\quad 0,\quad 0
$$
$$
-k_n,\quad k_n\xi_n,\quad k_n\xi_n,\quad 0
$$
$$
-m_{12},\quad m_{12}
$$
$$
-m_{1n},\quad m_{1n}
$$
$$
m_{n1},\quad -m_{n1}
$$
$$
m_{nn-1},\quad -m_{nn-1}
$$

(10)

From (4) and (7), the dynamic system of the Cournot model is given by

$$
(12) \quad
\begin{bmatrix}
dx_1/dt \\
dx_2/dt \\
\vdots \\
\vdots \\
dx_n/dt
\end{bmatrix}
=
\begin{bmatrix}
-k_1 & k_1\xi_1 & \cdots & k_1\xi_1 \\
k_2\xi_2 & -k_2 & \cdot & k_2\xi_2 \\
\vdots & & \ddots & \cdot \\
\vdots & \vdots & & \cdot \\
k_n\xi_n & k_n\xi_n & \cdots & -k_n
\end{bmatrix}
\begin{bmatrix}
x_1 \\
x_2 \\
\vdots \\
\vdots \\
x_n
\end{bmatrix}
+
\begin{bmatrix}
k\eta_1 \\
k_2\eta_2 \\
\vdots \\
\vdots \\
k_n\eta_n
\end{bmatrix}
$$

Call the coefficient matrix of (12) G. Since G has quasi-dominant negative diagonals under the same condition as H, that is, (11), the Cournot system is ensured stability under the same condition.

Consult the definition of ξ_i to rewrite as

$$(13) \quad 1/(n-1) > b/(d_i + 2b), \quad i=1,2,\cdots,n.$$

Considering the definition of profit function (3) one has,

$$\partial^2\pi_i/\partial x_i^2 = -(d_i + 2b) < 0,$$

$$\partial^2\pi_i/\partial x_j\partial x_i = -b < 0, \quad i \neq j,$$

$$\sum_{j \neq i} \partial^2\pi_i/\partial x_j\partial x_i = -(n-1)b < 0.$$

Hence the stability condition (13) is satisfied if the absolute rate of change of the marginal profit of the i-th firm with respect to change in its own output is greater than that with respect to simultaneous changes of all other firms' outputs for all i.

Let us consider implications of (13) in more detail. First, if n=2 (duopoly) we must have $d_1 > -b$, which holds, provided marginal cost of each firm is constant, increasing or decreasing absolutely in lesser degree than the absolute rate of change of price with respect to change in market supply. In the case of n=3 (triopoly), $d_i > 0$, meaning increasing marginal cost for each firm. In the case of $n \geq 4$, the stability condition (13) rewritten reads

$$(14) \quad \alpha < -b$$

where

$$\alpha \equiv \max_i \{-d_1/(n-3), -d_2/(n-3), \cdots, -d_n/(n-3)\}.$$

If $d_i = 0$ (constant marginal cost) for at least one i, (14) can not be met. We may see further from (13) that *ceteris paribus*, increase in the number of firms is more likely to violate the stability condition.

We have until now assumed that each firm has adaptive expectations on all rivals' outputs. As is clear from (4), however, what counts for a model of no product differentiation is each firm's expectation on the rest of industry output. This point can be introduced in an otherwise identical model as follows. Let $x^1 \equiv \sum_{j \neq i} x_j$ and x^{1e} be the i-th firm's expectation on x^1, the rest of industry output. The alternative dynamic system consists of (7), (15) and (16).

(7) $dx_i/dt = k_i(x_i^* - x_i),$ $i=1,2,\cdots,n,$

(15) $dx^{1e}/dt = m_i(x_i - x^{1e}),$ $m_i > 0,$ $i=1,2,\cdots,n,$

(16) $x_i^* = \eta_i + \xi_i x^{1e},$ $i=1,2,\cdots,n.$

A matrix representation of these equations is (17) on the next page.

Let a 2n x 2n matrix $A = [a_{ij}]$ be the coefficient matrix of (17), and define 2n positive coefficients q_i's as

$$q_i = \begin{cases} 1, & i=1,2,\cdots,n, \\ n-1, & i=n+1,n+2,\cdots,2n. \end{cases}$$

Under the stability condition for the Cournot system (11),

$$q_i|a_{ii}| > \sum_{j \neq i}^{2n} q_j|a_{ij}|,   i=1,2,\cdots,n$$

holds, while

$$q_i|a_{ii}| = \sum_{j \neq i}^{2n} q_j|a_{ij}|,   i=n+1,n+2,\cdots,2n$$

follows from definition of q_i's and A. Thus A has quasi-dominant negative diagonals, leading to stability of the solution for (17) under (11).

$$
\begin{pmatrix}
dx_1/dt \\
dx_2/dt \\
\vdots \\
dx_n/dt \\
dx^{1e}/dt \\
dx^{2e}/dt \\
\vdots \\
dx^{ne}/dt
\end{pmatrix}
=
\begin{pmatrix}
-k_1 & 0 & \cdots & 0 & k_1\xi_1 & 0 & \cdots & 0 \\
0 & -k_2 & \cdots & 0 & 0 & k_2\xi_2 & \cdots & 0 \\
\vdots & & \ddots & & & & & \vdots \\
0 & 0 & \cdots & -k_n & 0 & 0 & \cdots & k_n\xi_n \\
0 & m_1 & \cdots & m_1 & -m_1 & 0 & \cdots & 0 \\
m_2 & \cdots & m_2 & 0 & 0 & -m_2 & \cdots & 0 \\
\vdots & & & & & & \ddots & \vdots \\
m_n & \cdots & m_n & 0 & 0 & 0 & \cdots & -m_n
\end{pmatrix}
\begin{pmatrix}
x_1 \\
x_2 \\
\vdots \\
x_n \\
x^{1e} \\
x^{2e} \\
\vdots \\
x^{ne}
\end{pmatrix}
+
\begin{pmatrix}
k_1\eta_1 \\
k_2\eta_2 \\
\vdots \\
k_n\eta_n \\
0 \\
0 \\
\vdots \\
0
\end{pmatrix}
$$

(17)

7.2. Product Differentiation

The model in this section differs from that in Section 6.3 in that each firm's expectations on rivals' outputs are formed adaptively. Reproducing (1) through (4) in that section:

(1) $x_i = a_i - \sum_j b_{ij} p_j,$ $i=1,2,\cdots,n,$

(2) $C_i = e_i + c_i x_i + d_i x_i^2/2,$ $i=1,2,\cdots,n,$

$a_i > 0,\ b_{ii} > 0,\ b_{ij} < 0,$

(3) $\qquad\qquad\qquad\qquad\qquad i\neq j,\ j=1,2,\cdots,n,$

$c_i \geq 0,\ d_i \geq 0,\ e_i \geq 0,$

(4) $\pi_i(p) = p_i x_i - C_i,$ $i=1,2,\cdots,n.$

The expected profit maximizing price for the i-th firm is given by

(5) $p_i^* = \alpha_i - \sum_{j\neq i} r_{ij} p_{ij}^e,$ $i=1,2,\cdots,n,$

$\alpha_i \equiv (a_i + b_{ii} c_i + a_i b_{ii} d_i)/(2 + b_{ii} d_i) b_{ii},$ $i=1,2,\cdots,n,$

$r_{ij} \equiv (1 + b_{ii} d_i) b_{ij}/(2 + b_{ii} d_i) b_{ii},$ $i\neq j,\ j=1,2,\cdots,n.$

$p_{ij}^e,\ i \neq j,$ denotes the i-th firm's expectation on the j-th firm's price. In Section 6.3, we made a restrictive assumption that $p_{ij}^e = p_j^e$ for all $i \neq j.$ As expectations are of the adaptive type,

(6) $dp_{ij}^e/dt = m_{ij}(p_j - p_{ij}^e),$ $m_{ij} > 0,\ i\neq j,\ j=1,2,\cdots,n.$

The adjustment process for actual output is formulated as

(7) $dp_i/dt = k_i(p_i^* - p_i),$ $k_i > 0,\ i=1,2,\cdots,n.$

From the result in Section 6.3 we know that the Cournot oligopoly solution in a price adjusting model with product differentiation is stable if and only if

(8) $\sum_{j\neq i} r_{ij} > -1,$ $i=1,2,\cdots,n.$

From (5) - (7) the following system of differential equations is derived. (See equation (9) on the next page.)

Let F be the coefficient matrix of (9). F is partitioned as

$$
\begin{bmatrix} dp_1/dt \\ \vdots \\ dp_n/dt \\ dp_{12}^e/dt \\ \vdots \\ dp_{1n}^e/dt \\ \vdots \\ dp_{n1}^e/dt \\ dp_{nn-1}^e/dt \end{bmatrix}
=
\begin{bmatrix}
-k_1 & \cdots & 0 & -k_1 r_{12} & \cdots & -k_1 r_{1n} & \cdots & 0 & & & & 0 \\
 & & & & & 0 & & & & & & \\
0 & \cdots & -k_n & 0 & \cdots & 0 & \cdots & -k_n r_{n1} & \cdots & -k_n r_{nn-1} & & 0 \\
0\ m_{12} & \cdots & 0 & -m_{12} & \cdots & 0 & \cdots & 0 & & & & 0 \\
 & & & & & & & & & & & \\
0 & \cdots & m_{1n} & 0 & \cdots & -m_{1n} & \cdots & 0 & & & & 0 \\
 & & & & & & & & & & & \\
0 & \cdots & m_{n1} & 0 & & & \cdots & 0 & & & & 0 \\
0 & \cdots & m_{nn-1} & 0 & & & \cdots & 0 & & & & -m_{nn-1}
\end{bmatrix}
\begin{bmatrix} p_1 \\ \vdots \\ p_n \\ p_{12}^e \\ \vdots \\ p_{1n}^e \\ \vdots \\ p_{n1}^e \\ p_{nn-1}^e \end{bmatrix}
+
\begin{bmatrix} \alpha_1 k_1 \\ \vdots \\ \alpha_n k_n \\ 0 \\ \vdots \\ 0 \\ \vdots \\ 0 \\ 0 \end{bmatrix}
$$

(9)

$$F = \begin{bmatrix} F_{11} & F_{12} & \cdot\cdot & \cdot\cdot & F_{1,n+1} \\ F_{21} & F_{22} & \cdot\cdot & \cdot\cdot & F_{2,n+1} \\ \vdots & \vdots & & & \vdots \\ F_{n+1,1} & F_{n+1,2} & \cdot\cdot\cdot & \cdot\cdot & F_{n+1,n+1} \end{bmatrix} ,$$

where F: n^2 x n^2, F_{11}:   n x n,

F_{1j}:   n x (n-1),   j=2,$\cdots$,n+1,

F_{i1}:   (n-1) x n,   i=2,$\cdots$,n+1,

F_{ii}:   (n-1) x (n-1)   diagonal matrix, i=2,$\cdots$,n+1,

F_{ij}:   (n-1) x (n-1)   zero matrix, i$\neq$j, i,j=2,$\cdots$,n+1.

Stability of the solution for (9) under (8) can be established as follows. (See Okuguchi(1968) for a similar proof.) Define an n^2 x n^2 square matrix G

$$G = [g_{ij}] = F + \lambda I,$$

where λ is positive. By letting λ become sufficiently large, G can be made non-negative. Assume λ so chosen. By the Frobenius theorem on non-negative matrices, G has the Frobenius root $\rho(G)$, for which

(10) $\rho(G) \leq \sum_{j=1}^{n^2} g_{ij}$,   i=1,2,$\cdots$,$n^2$.

If (8), the stability condition for the Cournot system, is assumed,

(11) $\sum_{j=1}^{n^2} g_{ij} = \lambda - k_i - k_i \sum_{j \neq i} r_{ij}$

                  $< \lambda$,   i=1,2,$\cdots$,n.

The following equality is straightforward.

(12) $\sum_{j=1}^{n^2} g_{ij} = \lambda$,   i=n+1,n+2,$\cdots$,$n^2$.

Suppose $\rho(G) = \lambda$. Then,

$$|G - \rho(G)I| = |F| = 0$$

entailing indeterminacy of equilibrium for (9). Hence $\rho(G) \neq \lambda$. From

(11) and (12), $\rho(G) < \lambda$. Since characteristic roots of F equal those of G minus λ, real parts of characteristic roots of F are all negative. This completes proof of stability.

We now turn to a more restrictive case where expectations of all rivals on a firm's price are identical, thus $p_{ij}^e = p_j^e$ for all $i \neq j$. Equations (5) and (6) are replaced, respectively, by

$$(5') \qquad p_i^* = \alpha_i - \sum_{j \neq i} r_{ij} p_j^e, \qquad i=1,2,\cdots,n,$$

$$(6') \qquad dp_i^e/dt = m_i(p_i - p_i^e), \qquad m_i > 0, \; i=1,2,\cdots,n.$$

Combining (5'), (6') and (7),

$$(13) \qquad \begin{bmatrix} dp/dt \\ \\ dp^e/dt \end{bmatrix} = \begin{bmatrix} -K & -KR \\ \\ M & -M \end{bmatrix} \begin{bmatrix} p \\ \\ p^e \end{bmatrix} + \begin{bmatrix} K\alpha \\ \\ 0 \end{bmatrix},$$

where K and M are diagonal matrices whose (i, i) elements are k_i and m_i, and

$$R = \begin{bmatrix} 0 & r_{12} & \cdot & \cdot & \cdot & r_{1n} \\ r_{21} & 0 & \cdot & \cdot & \cdot & r_{2n} \\ \cdot & & \cdot & & & \cdot \\ & & & & & \vdots \\ r_{n1} & r_{n2} & \cdot & \cdot & \cdot & 0 \end{bmatrix},$$

$$\alpha = (\alpha_1, \alpha_2, \cdots, \alpha_n)',$$

$$p^e = (p_1^e, p_2^e, \cdots, p_n^e)'.$$

Stability of equilibrium for (13) under the stability condition for the Cournot system (8) can be proved by using similar arguments as we used in proving stability for (9). Tarr(1975), however, obtained a more general result showing stability-wise equivalence of the Cournot model and the adaptive system (13). Thus equilibrium for (13) is stable if and only if (8), the stability condition for the Cournot model is satisfied.

7.3. Mathematical Appendix

7.3.1. Some Theorems on Matrices with Quasi-Strictly-Dominant Diagonal Blocks

In Sections 7.1 and 7.2 we have encountered partitioned matrices such as H and F, respectively. We are here interested in stating without proofs some important theorems on partitioned matrices which have quasi-strictly-dominant diagonal blocks in a sense (which will soon be defined).

Let an n x n square matrix be partitioned into the following form.

$$(1) \qquad A = [a_{ij}] = \begin{bmatrix} A_{11} & A_{12} & \cdots & \cdots & A_{1N} \\ A_{21} & A_{22} & \cdots & \cdots & A_{2N} \\ \vdots & & \ddots & & \\ A_{N1} & A_{N2} & \cdots & & A_{NN} \end{bmatrix},$$

$$A_{IJ} = [a_{ij}^{IJ}], \qquad i=1,2,\cdots,n_I, \quad j=1,2,\cdots,n_J,$$

$$I,J=1,2,\cdots,N,$$

$$n_1 + n_2 + \cdots + n_N = n.$$

All diagonal blocks, that is, A_{II}'s are assumed to be non-singular. Define an N x N matrix $\Lambda(A)$ whose elements consist of norms of A_{IJ}'s:

$$(2) \qquad \Lambda(A) = \begin{bmatrix} \|A_{11}\| & \|A_{12}\| & \cdot & \|A_{1N}\| \\ \|A_{21}\| & \|A_{22}\| & \cdot \cdots \|A_{2N}\| \\ & & \ddots & \\ \|A_{N1}\| & \|A_{N2}\| & & \|A_{NN}\| \end{bmatrix}.$$

<u>Definition 7.3.1.1</u>: A is block indecomposable provided it has no permutation matrix P such as

$$P\Lambda(A)P^{-1} = \begin{bmatrix} A_1 & A_2 \\ 0 & A_3 \end{bmatrix},$$

where A_1 and A_3 are square matrices and A_2 is not necessarily a zero matrix.

Modifying and generalizing McKenzie(1960)'s definition of matrices with quasi-dominant diagonals (q.d.d.), we state

Definition 7.3.1.2: A partitioned matrix (1) has quasi-strictly-dominant diagonal blocks (q.s.d.d.b. for short) if for positive $d_I, I = 1,2,\cdots,N$,

$$(3) \quad d_I\| A_{II}^{-1}\|^{-1} > \sum_{J \neq I} d_J \| A_{IJ}\|, \qquad I=1,2,\cdots,N.$$

Remark 7.3.1.1: If A is block indecomposable and

$$(4) \qquad d_I\| A_{II}^{-1}\|^{-1} \geq \sum_{J \neq I} d_J\| A_{IJ}\|, \qquad I=1,2,\cdots,N,$$

where inequality is strict for at least one I, then A has q.s.d.d.b.. See Okuguchi(1976) and Y.Kimura(1970). Based on the foregoing definition, the following theorem is derived. For proofs of theorems given in this mathematical appendix, see Okuguchi(1975).

Theorem 7.3.1.1: A is non-singular if it has q.s.d.d.b..

Two further definitions must be given before stating further theorems.

Definition 7.3.1.3: Let $B = [b_{ij}]$ be a square matrix satisfying $b_{ii} > 0$ and $b_{ij} \leq 0$ for all i and j, $i \neq j$. B is called M-matrix (M for Minkowski) if for a diagonal matrix K whose diagonals are all positive, row sums of BK are all positive.

B is an M-matrix if and only if all principal minors of B are positive. See Ostrowski(1937).

Definition 7.3.1.4: Let

$$x = (x_1,x_2,\cdots,x_n), \quad x^+ = (|x_1|,|x_2|,\cdots,|x_n|).$$

A vector norm is is called an absolute norm provided norms for x and x^+ assume identical values, that is, $\| x \| = \| x^+\|$.

These additional definitions enable us to state

Theorem 7.3.1.2: Assume that A has q.s.d.d.b. and that the negatives of all diagonal blocks are M-matrices. Then real parts of all characteristic roots of A are negative, where a vector norm is understood to be absolute.

As is seen in the next subsection, this theorem has much relevance to stability of a system of difference equations whose constant coefficient matrix is partitionable.

Theorem 7.3.1.3: Let $m > 0$ and $a_{ii} \geq 0$ for all i. Then the largest absolute value of the characteristic root of A (that is, the spectral radius of A) is less than m provided $mI - A$ has q.s.d.d.b. with all diagonal blocks of $mI - A$ being M-matrices, where a vector norm is absolute.

This theorem and the immediate next have applicability to analysis of

stability of a system of differential equations whose coefficient matrix
is partitionable into blocks.

Theorem 7.3.1.4: The spectral radius of A is less than m if mI - A*A
has q.s.d.d.b. with all diagonal blocks being positive definite,
where A* is a conjugate transpose of A.

Theorem 7.3.1.5: The spectral radius of A is less than m provided

$$\sum_J \| A_{IJ} \| < m, \qquad I=1,2,\cdots,N$$

or when A is block indecomposable, if

$$\sum_J \| A_{IJ} \| \leq m, \qquad I=1,2,\cdots,N,$$

where inequality is strict for at least one I.

7.3.2. Some Applications

We are concerned here with applying some of the theorems stated in
the previous subsection. Consider first the coefficient matrix H of
(10) in Section 7.1 reproduced below.

$$H = \begin{bmatrix} H_{11} & & H_{1,n+1} \\ & \ddots & \\ & & \\ H_{n+1,1} & \cdots\cdots H_{n+1,n+1} \end{bmatrix},$$

H: n^2 x n^2, H_{11}: n x n,

H_{1j}: n x (n-1), j=2,$\cdots$,n,

H_{i1}: (n-1) x n, i=2,$\cdots$,n,

H_{ii}: (n-1) x (n-1) diagonal matrix whose diagonals are all
negative, i=2,$\cdots$,n+1.

H_{ij}: (n-1) x (n-1) zero matrix, i$\neq$j, i,j=2,$\cdots$,n+1.

It is clear that $-H_{ii}$'s are M-matrices. Let $\Lambda(H)$ be defined by

$$\Lambda(H) = \begin{bmatrix} \| H_{11} \| & & \| H_{1,n+1} \| \\ & \ddots & \vdots \\ & & \\ \| H_{n+1,1} \| & & \| H_{n+1,n+1} \| \end{bmatrix},$$

where $\| \cdot \|$ is to be taken as a (maximum) row sum norm.   Thus

$$\| H_{11} \| = \max_i k_i > 0, \quad \| H_{1j} \| = (n-1)k_j |\xi_j| > 0, \quad j=2,\cdots,n+1.$$

$$\| H_{11} \| = \max_{j \neq i} m_{ij} > 0, \quad i=2.\cdots,n+1,$$

$$\| H_{11} \| = \max_{j \neq i} m_{ij}, \quad i=2,\cdots,n+1,$$

$$\| H_{1j} \| = 0, \text{ otherwise.}$$

These properties of elements of $\Lambda(H)$ imply indecomposability of $\Lambda(H)$, that is, block indecomposability of H. To see this, suppose decomposability of $\Lambda(H)$. There must then exist two non-void subsets I and J of $N = \{1,2,\cdots,n+1\}$ such that

$$I \cup J = N, \quad I \cap J = \phi.$$

Let $1 \in I$. Since $\| H_{1j} \| \neq 0$, $j=1,2,\cdots,n+1$, we have $2,\cdots,n+1 \in I$. Thus $I = N$, which contradicts the definition of I (and J). Hence $1 \in J$. Next, let $2 \in I$. Since $\| H_{12} \| \neq 0$, we have $1 \in I$, leading to $I = N$ as above. Thus $2 \in J$. Similar arguments can be successively applied to lead to $3,\cdots,n+1 \in J$. Hence $J = N$, which is a contradiction. The upshot of these arguments is that $\Lambda(H)$ can not be decomposable.

Let us note further

$$\| H_{11}^{-1} \|^{-1} = \min_i k_i,$$

$$\| H_{ii}^{-1} \|^{-1} = \min_{j \neq i} m_{ij}, \quad i=2,\cdots,n+1,$$

and assume

(1)
$$k_i = k, \quad i=1,2,\cdots,n,$$
$$m_{ij} = m_i, \quad i \neq j, \; i,j=1,2,\cdots,n.$$

We can then show that H has q.s.d.d.b. provided

(2)
$$1 > (n-1)\sum_i |\xi_i| = -(n-1)\sum_i \xi_i$$

due to block indecomposability of H, proved above,

(3)
$$\| H_{11}^{-1} \|^{-1} > \sum_{j \neq 1} \| H_{1j} \| ,$$
$$\| H_{ii}^{-1} \|^{-1} = \sum_{j \neq i} \| H_{ij} \| , \quad i=2,\cdots,n+1,$$

and Remark 7.3.1.1. Theorem 7.3.1.2 then ensures negativity of real parts of all characteristic roots of H, hence stability. The stability condition (2) based on the restrictive assumption (1) is stronger than that which was derived in Section 7.1. The new stability condition (2) holds if

(4) $\xi_i > -1/n(n-1)$, $i=1,2,\cdots,n$.

We consider next the coefficient matrix of (9) in Section 7.2 whose partitioned form is seen to be

$$
F = \begin{bmatrix} F_{11} & \cdots & \cdots & \cdots & F_{1,n+1} \\ & \ddots & & & \\ \vdots & & \ddots & & \\ & & & \ddots & \\ F_{n+1,1} & & \cdots & \cdots & F_{n+1,n+1} \end{bmatrix},
$$

where F: $n^2 \times n^2$,

F_{11}: $n \times n$ diagonal matrix with diagonals all negative,

F_{1j}: $n \times (n-1)$, $j=2,\cdots,n+1$,

F_{ii}: $(n-1) \times (n-1)$ diagonal matrix whose diagonals are all negative, $i=2,\cdots,n+1$,

F_{ij}: $(n-1) \times (n-1)$ zero matrix, $i \neq j$, $i,j=2,\cdots,n+1$.

$-F_{ii}$s are M-matrices as is easily ascertained. Define $\Lambda(F)$ by

$$
\Lambda(F) = \begin{bmatrix} \| F_{11} \| & \cdot & \cdots & \cdots & \| F_{1,n+1} \| \\ & \ddots & & & \\ \vdots & & \ddots & & \vdots \\ & & & \ddots & \\ \| F_{n+1,1} \| & \cdots & \cdots & \| F_{n+1,n+1} \| \end{bmatrix} .
$$

Under an additional assumption,

(5)
$$k_i = k,   i=1,2,\cdots,n,$$
$$m_{ij} = m_i, \; i \neq j, \; i,j=1,2,\cdots,n,$$

F can be shown to have q.s.d.d.b. provided

(6) $1 > \sum\limits_{i} \sum\limits_{j \neq i} |r_{ij}| = -\sum\limits_{i} \sum\limits_{j \neq i} r_{ij}$.

This stability condition is stronger than (8) in Section 7.2 and holds if

$$\sum_{j \neq i} r_{ij} > -1/n, \qquad i = 1, 2, \cdots, n.$$

CHAPTER 8

UNKNOWN DEMAND FUNCTION AND STABILITY

8.1. Introduction

Except for Subsection 5.2.4 our analyses of oligopoly models have
been based on an assumption that each firm in an oligopolistic industry
has complete information of market demand function for its product as
well as for the market price and rivals' outputs in the past. In this
chapter we drop this assumption of complete information. Thus although each
firm does not in general have exact knowledge of the true market demand
function, it is assumed to have its own estimate of (or subjective)
market demand function which coupled with complete information of the
past market price enables it to estimate the rest of industry output in
the past. In Section 8.2 this information pattern is introduced into
the Cournot model involving no product differentiation. The Cournot
model with this information structure was first taken up by Y.Hosomatsu
(1969). Negishi and Okuguchi(1972) (see also Subsection 5.2.4) then
introduced the same information pattern into a Stackelberg-type leader-
leader model of duopoly. Hosomatsu, however, simplified his analysis
by using linear demand functions and quadratic cost functions. Our
analysis will be exempt from such restrictive specifications regarding
demand and cost functions. In Section 8.3 the Cournot assumption will
be replaced by adaptive expectations.

8.2. The Cournot Model with Unknown Market Demand Function

Our model in this section is formulated in terms of difference
equations. Throughout this section all functions are assumed to be as
many times differentiable as are required. Let there be n firms, and
let

(1) $p = f(\sum_i x_i),\ f' < 0$

be the true market demand function. Each firm being ignorant of this
demand function estimates it as

(2) $p = f_i(\sum_i x_i),\ f_i' < 0,\ i=1,2,\cdots,n,$

or if $g_i \equiv f_i^{-1},\ i=1,2,\cdots,n,$

(3) $\sum_i x_i = g_i(p),\ g_i' < 0,\ i=1,2,\cdots,n.$

Since the i-th firm has complete information of the market price and its own output in the past, its estimate of the rest of industry output in period t-1 defined by $y_i(t-1)$ is given by

(4) $\qquad y_i(t-1) = g_i(p(t-1)) - x_i(t-1), \qquad i=1,2,\cdots,n,$

where by virtue of (1)

(5) $\qquad p(t-1) = f(\sum_i x_i(t-1)), \qquad i=1,2,\cdots,n.$

Let each firm assume that the rest of industry output in period t will be the same as in period t-1 (that is, each firm has the Cournot assumption about the rest of industry output).

The i-th firm's expected profit in period t is defined to be

$$\pi_i^e(t) \equiv f_i(x_i(t) + g_i(f(\sum_j x_j(t-1))) - x_i(t-1))x_i(t) - C_i(x_i(t)),$$

$$i=1,\cdots,n.$$

Assuming instantaneous adjustment of actual output to a profit maximizing one and interior maximum, the i-th firm's output in period t has to satisfy,

$$f_i(x_i(t) + g_i(f(\sum_j x_j(t-1))) - x_i(t-1)) + f_i'(x_i(t)$$

(6)
$$+ g_i(f(\sum_j x_j(t-1))) - x_i(t-1))x_i(t) - C_i'(x_i(t)) = 0,$$

$$i=1,2,\cdots,n.$$

The second order condition is

(7) $\qquad 2f_i' + x_i(t)f_i'' - C_i'' < 0, \qquad i=1,2,\cdots,n.$

From (6),

(8) $\qquad x_i(t) = h_i(\sum_j x_j(t-1), x_i(t-1)), \qquad i=1,2,\cdots,n.$

This is a reaction function stating that the i-th firm's output in period t is a function of the total industry output and its own output in one earlier period. Define:

$$R^i \equiv \pi_i^e + C_i, \quad MR^i \equiv \partial R^i/\partial x_i(t), \quad MR_i^i \equiv \partial MR^i/\partial x_i(t),$$

$$MR_{i(-1)}^i \equiv \partial MR^i/\partial x_i(t-1), \quad MR_j^i \equiv \partial MR^i/\partial x_j(t-1), \qquad j\neq i.$$

From (6), expressions (9) and (10) follow.

(9) $\qquad \alpha_i \equiv \partial x_i(t)/\partial x_j(t-1) \equiv h_{ij} = -(f' + x_i(t)f_i''g_i'f')/(2f_i' + x_i(t)f_i''$

$$- C_i'') = MR_j^i/(MR_i^i - C_i''), \qquad j\neq i,$$

where $\quad MR_j^i \equiv \mu_i,\quad$ for all i, j≠i.

(10) $\quad \beta_i \equiv \partial x_i(t)/\partial x_i(t-1) \equiv h_{ii} = (f_i' + x_i(t)f_i'')/(2f'+ x_i(t)f_i'' - C_i'')$

$$= -MR_{i(-1)}^i/(MR_i^i - C_i'').$$

Applying the now familiar contraction mapping principle (see Mathematical Appendix 1.3.1) the Cournot oligopoly solution in a model with unknown market demand function and under no product differentiation is shown to exist, and is unique and globally stable provided

(11) $\quad |h_{ii}| + \sum_{j\neq i} |h_{ji}| < 1, \qquad i=1,2,\cdots,n,$

or rewritten

(12) $\quad |MR_{i(-1)}^i/(MR_i^i - C_i'')| + \sum_{j\neq i} |MR_j^j/(MR_j^i - C_j''| < 1, \qquad i=1,\cdots,n$

Taking into account $MR_j^i = \mu_i$ for all i, j≠i, it is immediately known that <u>ceteris paribus</u>, an increase in the number of firms is more likely to violate the stability condition (12).

To see what is implied by (12), consider a simple case where

$$p = a - b\sum_j x_j, \quad a,b > 0,$$

$$p = a_i - b_i \sum_j x_j, \quad a_i,b_i > 0, \qquad i=1,2,\cdots,n,$$

$$MC_i' = c_i = \text{const.}, \quad i=1,2,\cdots,n.$$

By simple calculations,

$$MR^1 = a_i - b_i(x_i(t) + (a_i/b_i - (a - b\sum_j x_j(t-1))/b_i)) - x_i(t-1))$$

$$- b_i x_i(t),$$

$$MR_{i(-1)}^i = b_i - b, \; MR_i^i = -2b_i, \; MR_j^i = -b, \qquad i\neq j.$$

The stability condition (12) then reduces to

(13) $\quad |b_i - b|/b_i + b \sum_{j\neq i} b_j^{-1} < 2, \qquad i=1,2,\cdots,n.$

Assume further that $b_i = b$ for all i, that is, each firm has exact knowledge of the slope of the market demand function. In this case (13) is satisfied if n = 2 (duopoly), which coincides with the result of Theocharis(1960) on the stability of the classical Cournot oligopoly solution. To be specifically noted here is the irrelevance of the intercept of true as well as subjective market demand function to stability as may be seen in (13).

8.3. Adaptive Expectations and Unknown Demand Function

In this section we shall extend our model of oligopoly with unknown market demand function by replacing the previously adopted Cournot assumption with adaptive expectations. From the beginning we shall assume that both the true market demand function and subjective ones are linear and that cost functions are quadratic.

Let the true market demand function and the i-th firm's subjective estimate of it be given, respectively, by

(1) $\qquad p = a - b\sum_i x_i, \quad a,b > 0$

(2) $\qquad p = a_i - b_i\sum_i x_i, \quad a_i,b_i > 0, \qquad i=1,2,\cdots,n.$

Define the rest of industry's actual output for the i-th firm by $q_i \equiv \sum_{j \neq i} x_j$. Given the market price and the i-th firm's actual output level in the immediately preceding period, the i-th firm estimates the value of q_i - let y_i be such an estimate-in the same period by utilizing its subjective demand function.

(3) $\qquad y_i = a_i/b_i - p/b_i - x_i$

$\qquad\qquad = (a_i - a)/b_i + (b - b_i)x_i/b_i + b\sum_{j \neq i} x_j/b_i.$

Let y_i^e be an expectation on y_i. Since expectations on the rest of industry output are assumed to be adaptively formed, y_i^e changes according to

(4) $\qquad dy_i^e/dt = m_i(y_i - y_i^e), \quad m_i > 0, \; i=1,2,\cdots,n.$

The i-th firm's quadratic cost function

$\qquad\qquad C_i = e_i + c_i x_i + dx_i^2/2, \qquad i=1,2,\cdots,n$

coupled with expected profit maximization gives rise to the first order and second order conditions below.

(5) $\qquad a_i - c_i - (2b_i + d_i)x_i - b_i y_i^e = 0, \qquad i=1,2,\cdots,n,$

(6) $\qquad 2b_i + d_i > 0, \qquad i=1,2,\cdots,n.$

The second order condition is met if d_i is non-negative, that is, the i-th firm's marginal cost is either increasing or constant. In our discussion that follows, we shall assume that this is the case.

Denoting the i-th firm's expected profit maximizing output by $\bar{x}_i$, (5) yields a reaction function

(7) $\qquad \bar{x}_i = (a_i - c_i)/(2b_i + d_i) - b_i y_i^e/(2b_i + d_i), \qquad i=1,2,\cdots,n.$

With lags in output adjustment,

(8) $dx_i/dt = k_i(\bar{x}_i - x_i),\ k_i > 0,\ i=1,2,\cdots,n.$

From (4), (7) and (8), equation (9) is derived by defining

$\gamma_i \equiv k_i(a_i - c_i)/(2b_i + d_i),\qquad i=1,2,\cdots,n.$

(9)

$$
\begin{bmatrix}
dx_1/dt \\
dx_2/dt \\
\vdots \\
dx_n/dt \\
dy_1^e/dt \\
dy_2^e/dt \\
\vdots \\
dy_n^e/dt
\end{bmatrix}
=
\begin{bmatrix}
-k_1 & 0 \cdots\cdots 0 & -k_1\alpha_1 & 0 \cdots & 0 \\
0 & -k_2 \cdots\cdot 0 & 0 & -k_2\alpha_2 \cdot & 0 \\
\vdots & \ddots & \vdots & & \vdots \\
0 & 0 \cdot\ -k_n & 0 & 0 \ \cdot\cdot -k_n\alpha_n \\
(\beta_1-1)m_1 & \beta_1 m_1\quad \beta_1 m_1 & -m_1 & 0 & 0 \\
\beta_2 m_2 & (\beta_2-1)m_2\cdot\cdot\beta_2 m_2 & 0 & -m_2 & \cdot\cdot 0 \\
\vdots & \ddots & & & \\
\beta_n m_n & \beta_n m_n\ (\dot\beta_n-1)m_n & 0 & 0 & -m_n
\end{bmatrix}
\begin{bmatrix}
x_1 \\
x_2 \\
\vdots \\
x_n \\
y_1^e \\
y_2^e \\
\vdots \\
y_n^e
\end{bmatrix}
+
\begin{bmatrix}
\gamma_1 \\
\gamma_2 \\
\vdots \\
\gamma_n \\
0 \\
0 \\
\vdots \\
0
\end{bmatrix}
$$

Let a $2n \times 2n$ matrix $A = [a_{ij}]$ be the coefficient matrix of (9), where $a_{ii} < 0$ for all i. Define $2n$ positive numbers, $q_1,q_2,\cdots,q_{2n}$ by

(10)

$q_j = 1,\ j = 1,2,\cdots,n.$

$q_{n+j} = (n - 1)\beta_j + |\beta_j - 1|,\qquad j=1,2,\cdots,n.$

We assume that q_i's satisfy

(11) $\alpha_j q_{n+j} < 1,\qquad j=1,2,\cdots,n$

From (10) and (11),

(12) $q_i|a_{ii}| > \sum_{j\neq i}^{2n} q_j|a_{ij}|,\qquad i=1,2,\cdots,n$

(13) $q_i|a_{ii}| = \sum_{j\neq i}^{2n} q_j|a_{ij}|,\qquad i=n+1,n+2,\cdots,2n,$

which show that A has quasi-dominant negative diagonals, proving stability of the solution for (9).

The stability condition (11) rewritten reads:

(14) $b/(2b + d_j) < 1/(n - 1)$ for $b_j = b$

(15) $b/(b_j + d_j) < 1/(n - 2)$ for $b_j > b$

(16) $b/(3b_j + d_j) < 1/n$ for $b_j < b$.

Now the interpretations of the stability condition. First, suppose $b_j = b$ for all j, thus all firms' estimates of the slope of the true market demand function are correct. This is the situation analyzed in Section 7.1. Inequality (14) is nothing but the stability condition thereby derived, and the interpretations thereof are directly applicable. Thus (14) is met if the absolute value of the rate of change of the marginal profit of each firm with respect to change in its own output is larger than that with respect to simultaneous changes of all rivals' outputs. In addition, other things being equal, the increase in the number of firms or b has a tendency to dissatisfy (14), while an increase in d_j is more likely to fulfill (14). Secondly, suppose that the dynamic system with a known market demand function is stable fulfilling (14). In the case of unknown market demand function, let at least one firm overestimate the slope of the market demand function, with $b_j > b$ for such firm(s). It is easily seen that the stability condition is satisfied for such firm(s). On the contrary, initial fulfillment of (14) is not sufficient for the validity of (16). We may further see that other things being equal, an increase in n or b is more likely to violate the stability condition, the reverse being true for an increase in d_j.

We shall next prove that the adaptive system with unknown demand function is always stable provided $k_i = k$ and $m_i = m$ for all i. Under our new assumption on adjustment coefficients, the coefficient matrix A simplifies to

$$A = \begin{bmatrix} -kI & -kG \\ -mF & -mI \end{bmatrix}$$

where $F = [f_{ij}]$, $f_{ii} = 1 - \beta_i$, $f_{ij} = \beta_i$, $j \neq i$ and G is a diagonal matrix whose (i, i) element is α_i.

Let λ be a characteristic root of a 2n x 2n matrix A and y be a characteristic vector of λ. From $Ay = \lambda y$, we get

(17) $-ky_i - \alpha_i ky_{n+i} = \lambda y_i$, $i=1,2,\cdots,n$,

(18) $m\beta_i y_1 + \cdots + m(\beta_i - 1)y_i + \cdots + m\beta_i y_n - my_{n+i} = \lambda y_{n+i}$, $i=1,\cdots,n$.

Combine (17) and (18) to get

(19) $-\beta_i \alpha_1 y_{n+1} - \cdots - (\beta_i - 1)\alpha_i y_{n+i} - \cdots - \beta_i \alpha_n y_{2n} = \mu y_{n+i}$, $i=1,\cdots,n$,

where $\mu \equiv (\lambda + m)(\lambda + k)/mk$ is a characteristic root of FG, hence of GF. Since $0 < \alpha_1 \le 1/2$ and $\beta_1 > 0$, it follows that GF satisfies all conditions of Hosomatsu(1969, Lemma (1.a), p.119)[1], hence all absolute values of characteristic roots of GF are less than unity, that is, $|\mu| < 1$. Consulting the definition of μ, non-negativity of the real part of λ is seen to be in contradiction to $|\mu| < 1$. Thus the real part of λ must be negative, completing out proof of invariable stability of (9) under $k_i = k$ and $m_i = m$ for all i.

------

1) Let an n x n matrix B be $B = [b_{ij}]$ with $b_{ij} = b_i$ for all $j \ne i$ and with all off-diagonal elements having the same signs. According to Hosomatsu, if $|b_{ii} - b_i| < 1$ and $b_i \le 0$ for all i, absolute values of all characteristic roots of B are less than one, if and only if,

$$1 + \sum_i b_i (1 + b_{ii} - b_i)^{-1} > 0.$$

See also Neudecker(1970, p.447). Though unnoticed by Hosomatsu, Neudecker demonstrated the remarkable fact that B's characteristic roots are all real.

CHAPTER 9

PROBABILITY MODELS

9.1. Probability Models with No Bayesian Learning

For duopoly models in Chapter 5 we assumed that each duopolist who
does not exactly know its rival's marginal cost function and the true
market demand function will perceive them subjectively. Oligopolists
in Chapter 8 were assumed to perceive subjectively only the unknown
market demand function. Notwithstanding their perceptions of the unknown
marginal cost and/or market demand function, firms in both chapters were
given no allowance for probabilistic consideration in choosing their
optimal policies. Once they formed subjective estimates of the relevant
function(s), no revision of estimates occurred. Subjective estimates
were rather of a definite nature and probability distribution of the
estimates was beyond consideration.

Following R.M.Cyert and M.H.DeGroot(1970b), let us first see how
firms' probabilistic expectations of rivals' behaviors and a parameter
of the unknown market demand function can be introduced into a simple
single period duopoly model of simultaneous choice. Let the unknown
market demand function be

$$p = a - b(x_1 + x_2),$$

and assume no production cost as well as perfect knowledge of the slope
of the market demand function for both firms. The i-th firm's profit is

$$\pi_i = x_i(a - b(x_i + x_j)) \qquad i \neq j, \ i,j=1,2.$$

Let E_1 and E_2 be expectation operators for the first and second firms,
respectively. Hence,

(1) $\qquad E_i(\pi_i) = x_i E_i(a) - bx_i^2 - bx_i E_i(x_j), \qquad i \neq j, \ i,j=1,2,$

which is maximized when

(2) $\qquad x_i = (E_i(a) - bE_i(x_j))/2b, \qquad i \neq j, \ i,j=1,2.$

Hence

(3) $\qquad x_i = E_i(a)/2b - E_i E_j(a)/4b + E_i E_j(x_i)/4, \qquad i \neq j, \ i,j=1,2,$

where $E_i E_j(\cdot)$ denotes the i-th firm's expectations as to the j-th firm's
expectations for a random variable in the parenthesis. From (2) and (3)
we get

$$x_i = E_i(a)/2b - E_iE_j(a)/4b + E_iE_jE_i(a)/8b - E_iE_jE_i(x_j)/8$$

(4)
$$= E_i(a)/2b - E_iE_j(a)/4b + E_iE_jE_i(a)/8b$$

$$- E_iE_jE_iE_j(a)/16b + E_iE_jE_iE_j(x_i)/16, \qquad i \neq j, \ i,j=1,2,$$

to which substitutions are possible an infinite number of times.
Complicated expressions in (4) show "I think that he thinks that I think
..." difficulties of an infinite chain of mutual expectations in a
single period duopoly model of simultaneous decisions of two firms.
The Cournot equilibrium will emerge when both firms are assumed to know
correctly the value of a and when at the same time each firm is assumed
to be aware of this fact. For under these assumptions, (4) results in

$$x_i = a/2b - a/4b + a/8b - a/16b + \cdots$$

(5)
$$= a/3b, \qquad i=1,2$$

which is nothing but the optimal outputs at the Cournot duopoly equilibrium.

Probabilistic elements regarding firms' expectations on rivals'
behaviors can be introduced into dynamic oligopoly models based on and
extending I.Horowitz(1970a) in the following ways. For oligopoly under
no product differentiation, let

$$c_i = c_i(x_i), \qquad i=1,2,\cdots,n$$

be the i-th firm's cost function and

$$p = f(\sum_i x_i)$$

be the market demand function of which all firms are assumed to be fully
aware. Define

$$x^i \equiv \sum_{j \neq 1} x_j, \qquad i=1,2,\cdots,n$$

which represents the rest of the industry output for the i-th firm; let
x^{ie} be its expectation and

$$y_i \equiv x^{ie}.$$

It is assumed that the expected value of the rest of the industry output
for the i-th firm in period t equals the actual value of the rest of the
industry output in period t-1, hence

$$E(y_i(t)) \equiv \mu_i(t) = x^i(t-1), \qquad i=1,2,\cdots,n$$

where $y_i(t)$ is a random variable for the i-th firm, and E represents
the expectation operator. Probability density function for $y_i(t)$ will

be denoted

(6) $\qquad g_{i,t}(y_i(t); x^1(t-1))$, $\qquad i=1,2,\cdots,n$

In the following, $y_i(t)$ will be simply written y_i for notational convenience. Define further

$$p_{i,t} \equiv f(x_i(t) + y_i), \quad f' < 0,$$

$$\pi_{i,t} \equiv x_i(t)p_{i,t} - C_i(x_i(t))$$

$$U_i \equiv U_i(\pi_{i,t}), \text{ the i-th firm's utility function,}$$

$$1/h_{i,t} \equiv E(y_i - \mu_i(t))^2, \text{ variance of } y_i,$$

all for $i=1,2,\cdots,n$. Suppose that the i-th firms is a von Neumann-Morgenstern expected utility maximizer. The first and second order conditions for expected utility maximizations are:

(7) $\qquad dE(U_i)/dx_i(t) = \int_{y_i} U_i' \cdot (p_{i,t} + x_i(t)p_{i,t}' - C_i')g_{i,t}(y_i)dy_i$

$$= E(U_i')(E(p_{i,t}) + x_i(t)E(p_{i,t}') - C_i' +$$

$$Cov(U_i', p_{i,t}) + x_i(t)Cov(U_i', p_{i,t}') = 0,$$

$$i=1,2,\cdots,n,$$

(8) $\qquad d^2E(U_i)/d^2x_i(t) = \int_{y_i} U_i'' \cdot (p_{i,t} + x_i(t)p_{i,t}' - C_i')^2 g_{i,t}(y_i)dy_i$

$$+ \int_{y_i} U_i' \cdot (2p_{i,t}' + x_i(t)p_{i,t}'' - C_i'')g_{i,t}(y_i)dy_i < 0,$$

$$i=1,2,\cdots,n.$$

The second order condition is satisfied provided $U_i'' \leq 0$ (the i-th firm is risk-neutral or risk-averse) and $\partial^2 \pi_{i,t}/\partial x_i^2(t) < 0$ for arbitrary y_i. Assume

$$p_{i,t}^{(k)} \equiv \partial^k p_{i,t}/\partial(x_i(t) + y_i)^k = 0 \text{ for } k > 3;$$

$g_{i,t}(y_i)$ is symmetric, and let

$$p_i^{(k)} \equiv p_{i,t}^{(k)} \text{ evaluated at } x_i(t) + \mu_i(t), k=0,1,\cdots,$$

The first order condition (7) can be rearranged to read

(9) $\qquad p_i + p_i'x_i(t) + (p_i'' + x_i(t)p_i''')/2h_{i,t}$

$$+ (Cov(U_i', p_{i,t}) + x_i(t)Cov(U_i', p_{i,t}'))/E(U_i') = C_i'$$

$$i=1,2,\cdots,n,$$

which defines implicitly the reaction function for the i-th firm. In the original Cournot model, $h_{i,t} = \infty$ and two covariances in the above expressions vanish. Even if $h_{i,t} \neq \infty$, the original Cournot model will emerge provided U_i is linear (that is, the i-th firm is risk-neutral) and $p_i'' + x_i(t)p_i''' = 0$, which is true for a linear market demand function.

Let $x_i^N(t)$, $x_i^A(t)$ and $x_i^P(t)$ be the optimal outputs for the i-th firm when it is risk-neutral, risk-averse and risk-preferring, respectively. Assuming $U_i' > 0$ and $p_{i,t}'' \leq 0$, we can infer from (7) and (8) that

(10) $\qquad x_i^A(t) < x_i^N(t) < x_i^P(t), \qquad i=1,2,\cdots,n.$

To see this we have only to take into account:

$$\text{Cov}(U_i'(\pi_{i,t}),p_{i,t}) \begin{cases} < 0, \text{ if the i-th firm is risk-averse} \\ = 0, \text{ if it is risk-neutral} \\ > 0, \text{ if it is risk-preferring} \end{cases}$$

$$\text{Cov}(U_i'(\pi_{i,t}),p_{i,t}') \begin{cases} \leq 0, \text{ if the i-th firm is risk-averse} \\ = 0, \text{ if it is risk-neutral} \\ \geq 0, \text{ if it is risk-preferring.} \end{cases}$$

A similar result as in (10) has been established for a competitive firm facing product price uncertainty by J.J.McCall(1967), M.Rothschild and J.E.Stiglitz(1970, 1971), K.Okuguchi(1970) and A.Sandmo(1971).

As for the stability of the equilibrium in (9), nothing definite can be asserted in general. However, the stability condition may be derived without much difficulty provided the last term on the left-hand side of (9) vanishes and $1/h_{i,t}$ is assumed to equal $\mu_i(t) = x^j(t-1) = \sum_{j \neq i} x_j(t-1)$.

Horowitz analyzed a mechanism of a firm's Bayesian learning regarding probability density functions of rival's possible output for a simple duopoly model.

According to D.Tarr(1972), probability density function of the rest of the industry output for the i-th firm, h_i, depends on all firms' actual outputs in the last m periods[1], hence

_____

1) Since learning is excluded in the Tarr model, h_i does not explicitly depend on t in contrast to Horowitz. Let more generally

(11') $\qquad h_{i,t} = h_{i,t}(y_i; x(t-1),x(t-2),\cdots,x(t-m)),$

then (6) may be taken as a special case of (11').

(11) $h_i \equiv h_i(y_i; x(t-1), x(t-2), \cdots, x(t-m))$, $i=1,2,\cdots,m$,

where

$$x = (x_1, x_2, \cdots, x_n)'.$$

If the i-th firm is an expected utility maximizer[1], the first and second order conditions for expected utility maximization are given by (7) and (8) with $g_{i,t}$ replaced by h_i. Under the assumption of explicit solvability, the first order condition yields

(12) $x_i(t) \equiv H_i(x(t-1), x(t-2), \cdots, x(t-m))$, $i=1,2,\cdots,n$,

which represents the reaction function for the i-th firm. To analyze the stability of the equilibrium in (12), introduce a transformation

(13) $w(t) = (w_1(t), w_2(t), \cdots, w_{mn}(t))'$

$$\equiv (x(t), x(t-1), \cdots, x(t-m+1))',$$

and define

$$a_{ij} \equiv \partial H_i / \partial w_j(t-1), \qquad i=1,2,\cdots,n$$
$$j=1,2,\cdots,mn$$

Let $x^* = (x_1^*, x_2^*, \cdots, x_n^*)'$ be the equilibrium for (12) and a_{ij}^*'s be a_{ij}'s evaluated there. Expand (12) around the equilibrium and introduce an mn-vector Z.

$$Z = (Z_1, Z_2, \cdots, Z_{mn})'$$

where

$$Z_{(k-1)n+i} \equiv y_{(k-1)n+i} - x_i^*, \qquad k=1,2,\cdots,m$$
$$i=1,2,\cdots,n.$$

We then have

(14) $Z(t) = AZ(t-1)$,

where an mn x mn matrix A defined by

---

1) Tarr's analysis is restricted to a risk-neutrality case.

$$
A = \begin{bmatrix}
a^*_{11} & \cdots & a^*_{1n} & a^*_{1,n+1} & \cdots & a^*_{1,2n} & \cdots & a_{1,(m-1)n+1} & & a_{1,mn} \\
\vdots & & \vdots & \vdots & & \vdots & & \vdots & & \vdots \\
a^*_{n1} & & a^*_{nn} & a_{n,n+1} & & a_{n,2n} & & a_{n,(m-1)n+1} & & a_{n,mn} \\
1 & \cdots & 0 & 0 & \cdots & 0 & & 0 & \cdots & 0 \\
\vdots & \ddots & \vdots & \vdots & & \vdots & & \vdots & & \vdots \\
0 & \cdots & 1 & 0 & \cdots & 0 & & 0 & \cdots & 0 \\
0 & \cdots & 0 & 1 & \cdots & 0 & & 0 & \cdots & 0 \\
& & & \vdots & \ddots & \vdots & & \vdots & & \vdots \\
0 & \cdots & 0 & 0 & \cdots & 1 & & 0 & \cdots & 0 \\
0 & \cdots & 0 & 0 & \cdots & 0 & & 1 & \cdots & 0 \\
\vdots & & \vdots & \vdots & & \vdots & & & \ddots & \vdots \\
0 & \cdots & 0 & 0 & \cdots & 0 & & 0 & \cdots & 1
\end{bmatrix}
$$

can be partitioned like

$$
A = [A_{IJ}] = \begin{bmatrix}
A_{11} & A_{12} & \cdots & \cdots & A_{1m} \\
I & 0 & & & 0 \\
0 & I & \cdots & \cdots & 0 \\
\vdots & \vdots & & & \vdots \\
0 & 0 & & & I \; 0
\end{bmatrix}
$$

and

$$
A_{1,J} = \begin{bmatrix}
a^*_{1,(J-1)n+1} & & a^*_{1,Jn} \\
a^*_{2,(J-1)n+1} & \cdots & a^*_{2,Jn} \\
\vdots & & \vdots \\
a^*_{n,(J-1)n+1} & & a^*_{n,Jn}
\end{bmatrix}
\qquad J = 1,2,\cdots,n.
$$

Assume that all A_{IJ}'s have at least one non-zero element, and let $\| \; \|$ be any matrix norm such that $\|I\| = 1$. By virtue of Theorem 7.3.1.5, the moduli of all characteristic roots of A are less than one, leading to the local stability of the equilibrium for (12), provided the following inequality holds.

$$(15) \qquad \sum_{J=1}^{m} \| A_{IJ} \| < 1.$$

The stability condition is satisfied[1] when

$$(16) \qquad \sum_{J=1}^{m} \max_{i=1,2,\cdots,n} \sum_{j=(J-1)n+1}^{Jn} a^{*}_{ij} < 1$$

From this one can conclude that the equilibrium is more likely to be locally stable the less sensitive change in each firm's output is with respect to change in output of any firm in the relevant past period. The stability condition (15) is closely related to the stability condition for general linear distributed lag models. See J.Conlisk(1973) and Okuguchi(1975).

9.2. Bayesian Learning in Duopoly Models

The Bayesian learning mechanism was introduced by Horowitz(1970a) into a duopoly model where duopolists maximized their expected utility in each period and where the variance of y_i was assumed to equal its expected value. Cyert and DeGroot(1970b, 1971, 1973) developed multi-period expected profit maximizing duopoly models involving Bayesian learnings. Earlier J.W.Friedman(1968) constructed a multiperiod duopoly model where duopolists were assumed to choose simultaneously their optimal outputs in each period so as to maximize their discounted profits over finite or infinite periods. If the planning period is finite and simplifying assumptions are introduced, duopolists' optimal outputs in each period in the Friedman model coincides with the original Cournot equilibrium outputs.[2]

In order to remove this trivial and uninteresting property of the multiperiod duopoly model, Cyert and DeGroot(1970a, 1973) introduced a rather unnatural assumption that two firms choose their outputs in alternate periods.

To see how Bayesian learning will take place in a context of multi-period duopoly, consider first a single period leader-follower type

---

1) Take $\| \; \|$ to be the row sum norm. Note that the stability condition (16) slightly differs from the one obtained by Tarr.
2) See Cyert and DeGroot(1970a, pp.413 - 415) for detailed argument on this point.

duopoly model of Cyert and DeGroot(1970b) where the market demand function which is assumed to be perfectly known by both firms is given by

(1) $\qquad p = a - b(x_1 + x_2), \quad a,b > 0.$[1]

It is assumed that production cost is zero for the first firm designated as the leader and that the leader chooses its optimal output before the second firm called the follower. After learning the value of x_1, the follower chooses its output according to

(2) $\qquad x_2 = x_1 + \theta + \epsilon.$

Here θ is a constant which is fixed by the follower prior to its knowledge of the value of x_1, and ϵ is a random variable perfectly known by both firms. Let $m = E(\theta)$. The leader's expected profit is then given by

(3) $\qquad E(\pi_1) = ax_1 - 2bx_1^2 - bmx_1,$

which is maximized when

(4) $\qquad x_1 = (a - bm)/4b.$

The corresponding expected profit is

(5) $\qquad E(\pi_1) = (a - bm)^2/8b.$

Consider now a T-period model. Let $x_i(t)$, $i = 1, 2$, be the i-th firm's output in period $t = 1, 2, \cdots, T$, and assume no cost of production for the leader. The leader's profit in period t, $\pi_1(t)$, is

(6) $\qquad \pi_1(t) = ax_1(t) - bx_1^2(t) - bx_1(t)x_2(t), \qquad t=1,2,\cdots,T.$

The follower is assumed to act on the basis of a similar equation as (2) in each period. Hence,

(7) $\qquad x_2(t) = x_1(t) + \theta + \epsilon_t, \qquad t=1,2,\cdots,T,$

where θ is assumed to remain constant over time and ϵ_t's are independent, identically distributed, normal random variables with mean 0 and known precision (which is the reciprocal of the variance)τ. The leader is assumed to maximize its expected profit over T periods

---

[1] In the formulation of Cyert and DeGroot, the market demand function is

(1') $\qquad p = a - bx_1 - cx_2,$

where b and c are not necessarily identical. Though (1') may be reasonable for a model with product differentiation, its rationale is rather dubious when the product is not differentiated.

(8) $E(\pi_1(1) + \pi_1(2) + \cdots + \pi_1(T))$.

The problem to be solved here is to find the leader's optimal sequence of outputs which maximizes (8).

The optimal outputs can be found by backward induction. Suppose that the first T-1 periods are over and that the leader is faced with the choice of optimal output in the final T-th period. Since profits for the first T-1 periods have already been determined and are unchangeable, the only thing left for the leader to do is to maximize its expected profit in the final period. Let the prior distribution of θ for the leader at the beginning of the T-th period be normal with mean m' and precision h'. From (4) and (5), the optimal output and maximum expected profit in the T-th period are

(9) $x_1(T) = (a - bm')/4b$,

(10) $E(\pi_1(T)) = (a - bm')^2/8b$.

Moving one period backward, suppose that the first T-2 periods are over and that the leader is to choose its optimal output in the (T-1)-th period. Let its prior distribution for θ at the beginning of the (T-1)-th period be normal with mean m and precision h. The leader now faces with maximization of expected profit over the last two periods,

(11) $E(\pi_1(T-1)) + E(\pi_1(T))$.

Before proceeding further let us·cite[1)]
Theorem 9.2.1: Suppose that $x_1, x_2, \cdots, x_n$ are random samples from a normal distribution with an unknown mean μ and a known precision h. Let the prior distribution of μ be normal with mean m and precision τ. Then the posterior distribution of μ when $x_1 = x_1^*, x_2 = x_2^*, \cdots, x_n = x_n^*$ is normal with mean m' and precision τ', where

$m' = (\tau m + nh\bar{x}^*)/(\tau + nh)$,

$\tau' = \tau + nh$,

$\bar{x}^* = \sum_i x_i^*/n$.

By virtue of this theorem,

(12) $m' = (hm + \tau(x_2(T-1) - x_1(T-1)))/h'$, $h' = h + \tau$.

From (10) and (12),

_____

1) See DeGroot, M.H., <u>Optimal Statistical Decisions</u>, McGrawHill, New York, 1970, p.167.

(13) $E(\pi_1(T)) = (a - bm')^2/8b$

$= (a - b(hm + \tau(x_2(T-1) - x_1(T-1)))/(h + \tau))^2/8b,$

where by virtue of (7), $x_2(T-1)$ is a random variable for any given value of $x_1(T-1)$. In fact $x_2(T-1)$ is normal with mean $x_1(T-1) + m$ and precision $h\tau/(h + \tau)$. Taking expectation of (13) with respect to $x_2(T-1)$ for given $x_1(T-1)$, we get

(14) $E(\pi_1(T)) = ((a - bm)^2 + \tau b^2/h(h + \tau))/8b,$

which is independent of $x_1(T-1)$. In order to maximize (11) the leader therefore chooses $x_1(T-1)$ to maximize only $E(\pi_1(T-1))$. Hence the leader's optimal output and expected profit in the $(T-1)$-th period are given by

(15) $x_1(T-1) = (a - bm)/4b$

and

(16) $E(\pi_1(T-1)) = (a - bm)^2/8b,$

respectively. From (14) and (16), the leader's maximum expected profit over the last two periods equals

(17) $E(\pi_1(T-1)) + E(\pi_1(T)) = (a - bm)^2/4b + \tau b/8h(h + \tau).$

Similar arguments can be successively applied to the determination of the leader's optimal outputs and expected profits in earlier stages and the maximization problem can be completely solved.

The method of backward induction can also be applied to derive the leader's optimal outputs when the follower is assumed to choose its output in each period according to

(18) $x_2(t) = \theta x_1(t) + \epsilon_t,$ $t=1,2,\cdots,T,$

where θ is fixed positive number chosen by the follower and ϵ_t's are independent, identically distributed normal random variables with mean o and known precision τ.

Bayesian learning was introduced also in Cyert and DeGroot(1971) to reveal a condition consistent with the kink in the demand function in a multiperiod expected profit maximizing duopoly model. Furthermore Cyert and DeGroot(1973) were concerned with learning in a multiperiod duopoly model with an alternate choice of and cooperation between firms. The interested reader is advised to turn directly to Cyert and DeGroot's works.

REFERENCES

Arrow,K.J. and G.Debreu, "Existence of an Equilibrium for a Competitive
    Economy", Econometrica, 22 (1954)

Arrow,K.J. and M.Nerlove, "A Note on Expectations and Stability", Econo-
    metrica, 26 (1958)

Baumol,W., "On the Theory of Oligopoly", Economica, N.S. 25 (1958)

Baumol,W., Business Behavior, Value and Growth, MacMillan, New York,
    1959

Beckmann,M.J., "Spatial Cournot Oligopoly", Papers of the Regional
    Science Association, 28 (1972)

Bishop,R.L., "The Stability of the Cournot Oligopoly Solution: Further
    Comment", Review of Economic Studies, 29 (1962)

Brock,W.A. and J.A.Scheinkman, "Some Results on Global Asymptotic
    Stability of Difference Equations", Journal of Economic Theory,
    10 (1975)

Burger,E., Introduction to the Theory of Games, Prentice-Hall, New Jersey,
    1963 (Original German edition, 1959)

Cagan,P., "Monetary Dynamics of Hyper-Inflation", in M.Friedman (ed.),
    Studies in Quantity Theory of Money, The University of Chicago
    Press, Chicago, 1956

Chen,C.I. and J.B.Cruz Jr., "Stackelberg Solution for Two-Person Games
    with Biased Information Patterns", IEEE Transactions on Automatic
    Control, AC-17 (1972)

Conlisk,J., "Quick Stability Checks and Matrix Norms", Economica, 40
    (1973)

Cournot,A., Researches into the Mathematical Principles of the Theory
    of Wealth, 1897 (Original French edition, 1838)

Cyert,R.M. and M.H.DeGroot, "Multiperiod Decision Models with Alternating
    Choice as a Solution to the Duopoly Problem", Quarterly Journal of
    Economics, 84 (1970a)

Cyert,R.M. and M.H.DeGroot, "Bayesian Analysis and Duopoly Theory",
    Journal of Political Economy, 78 (1970b)

Cyert,R.M. and M.H.DeGroot, "Interfirm Learning and the Kinked Demand
    Curve", Journal of Economic Theory, 3 (1971)

Cyert,R.M. and M.H.DeGroot, "An Analysis of Cooperation and Learning in
    a Duopoly Context", American Economic Review, 63 (1973)

Debreu,G. and I.N.Herstein, "Nonnegative Square Matrices", Econometrica,
    21 (1953)

Debreu,G. and H.Scarf, "A Limit Theorem on the Core of an Economy",
    International Economic Review, 4 (1963)

Enthoven,A.C. and K.J.Arrow, "A Theorem on Expectations and the Stability
    of Equilibrium", Econometrica, 24 (1956)
Fellner,W., Competition among the Few, Sentry Press, New York, 1949
Fisher,F.M., "The Stability of the Cournot Oligopoly Solution: The
    Effects of the Speeds of Adjustment and Increasing Returns",
    Review of Economic Studies, 28 (1961)
Fisher,F.M., "Quasi-Competitive Price Adjustment by Individual Firms:
    A Preliminary Paper", Journal of Economic Theory, 2 (1970)
Formby,J.P., "On Revenue Maximizing Duopoly", Journal of Industrial
    Economics, 21 (1973)
Frank,C.R.Jr. and R.E.Quandt, "On the Existence of Cournot Equilibrium",
    International Economic Review, 5 (1963)
Frank,C.R.Jr., "Entry in a Cournot Market", Review of Economic Studies,
    32 (1965)
Friedman,J.W., "Reaction Functions and the Theory of Duopoly", Review
    of Economic Studies, 35 (1968)
Friedman,J.W., "On the Structure of Oligopoly Models with Product
    Differentiation", in H.Sauermann (ed.), Contributions to Experi-
    mental Economics, J.C.B.Mohr, Tubingen, 1972
Friedman,J.W., Oligopoly Theory, manuscript, 1975
Frish,R., "Monopole-Polypole - La Notion de la Force dans l'Economie",
    Nationaløkonomik Tidsskrift, 71 (1933), English translation in
    E.Henderson (ed.), International Economic Papers, No.1, MacMillan,
    London, 1951
Gabscewicz,J.J. and J.Vial, "Oligopoly 'a la Cournot' in General Equilib-
    rium Analysis", Journal of Economic Theory, 4 (1972)
Gale,D., "The Law of Supply and Demand", Mathematica Scandinavica, 3
    (1955)
Gale,D., "On Equilibrium for a Multi-Sector Model of Income Propagation",
    International Economic Rewiew, 5 (1964)
Gale,D. and H.Nikaido, "The Jacobian Matrix and Global Univalence of
    Mappings", Mathematische Annalen, 159 (1965)
Hadar,J., "A Note on Dominant Diagonals in Stability Analysis",
    Econometrica, 33 (1965)
Hadar,J., "Stability of Oligopoly with Product Differentiation", Review
    of Economic Studies, 33 (1966)
Hadar,J., "On Expectations and Stability", Behavioral Science, 13 (1968)
Hahn,F.H., "The Stability of the Cournot Oligopoly Solution", Review
    of Economic Studies, 29 (1962)
Hawkins,C.J., "On the Sales Revenue Maximization Hypothesis", Journal
    of Industrial Economics, 8 (1970)

Henderson,J.M. and R.E.Quandt, Microeconomic Theory, McGraw-Hill, New
     York, 1958

Horowitz,I., "Nondogmatic Conjectures in a Cournot Market", Western
     Economic Journal, 8 (1970a)

Horowitz,I., Decision Making and the Theory of the Firm, Holt, Rinehart
     and Winston, New York, 1970b

Hosomatsu,Y., "A Note on the Stability Conditions in Cournot's Dynamic
     Market Solution", Review of Economic Studies, 36 (1969)

Inada,K., "Factor Intensity and Stolper-Samuelson Condition", Econo-
     metrica, 39 (1971)

Iwata,G., Measurement of Conjectural Variations in Oligopoly", Econo-
     metrica, 42 (1974)

Kakutani,S., "A Generalization of Brouwer's Fixed Point Theorem", Duke
     Mathematical Journal, 8 (1941)

Kamerschen,D.R. and P.E.Smith, "Duopoly Models and Global Stability",
     Economic Studies Quarterly, 22 (1971)

Kamien,M.I. and N.L.Schwarz, "Cournot Oligopoly with Uncertain Entry",
     Review of Economic Studies, 42 (1975)

Kimura,Y., "On the Roles of Matrices with Semi-Dominant Diagonals in
     Economic Theory", Oikonomica, 7 (1970) (in Japanese)

Kolmogoroff,A. and S.V.Fomin, Elements of Theory of Functions and Func-
     tional Analysis, Graylock Press, Rochester, N.Y., 1957 (Original
     Russian edition 1954)

Marschak,T. and R.Selten, General Equilibrium with Price-Making Firms,
     Vol. 91, Lecture Notes in Economics and Mathematical Systems,
     Springer-Verlag, Berlin, Heidelberg and New York, 1974

Marshall,A., Principles of Economics, MacMillan, London, 1890

Mayberry,J.P., J.F.Nash and M.Shubik, "A Comparison of Treatments of a
     Duopoly Situation", Econometrica, 21 (1953)

McCall,J.J., "Competitive Production for Constant Risk Utility Functions",
     Review of Economic Studies, 34 (1967)

McKenzie,L.W., Matrices with Dominant Diagonals and Economic Theory",
     in K.J.Arrow et.al. (eds.), Mathematical Methods in the Social
     Sciences, Stanford University Press, Stanford, 1960

McManus,M. and R.E.Quandt, "Comments on the Stability of the Cournot
     Oligopoly Model", Review of Economic Studies, 28 (1961)

McManus,M., "Dynamic Cournot-Type Oligopoly Models: A Correction",
     Review of Economic Studies, 29 (1962a)

McManus,M., "Numbers and Size in Cournot Oligopoly", Yorkshire Bulletin
     of Social and Economic Research, 14 (1962b)

McManus,M., "Equilibrium, Numbers and Size in Cournot Oligopoly",
     Yorkshire Bulletin of Social and Economic Research, 16 (1964)

Morishima,M., <u>Equilibrium</u>, <u>Stability</u> <u>and</u> <u>Growth</u>, Oxford University
     Press, London, 1964
Morishima,M., <u>Theory</u> <u>of</u> <u>Economic</u> <u>Growth</u>, Oxford University Press,
     London, 1969
Negishi,T., "Monopolistic Competition and General Equilibrium", <u>Review</u>
     <u>of</u> <u>Economic</u> <u>Studies</u>, 28 (1961)
Negishi,T., "The Stability of Exchange and Adaptive Expectations",
     <u>International</u> <u>Economic</u> <u>Review</u>, 5 (1964)
Negishi,T. and K.Okuguchi, "A Model of Duopoly with Stackelberg Equilib-
     rirm", <u>Zeitscrift</u> <u>für</u> <u>Nationalökonomie</u>, 32 (1972)
Nerlove,M., <u>The</u> <u>Dynamics</u> <u>of</u> <u>Supply</u>: Estimation <u>of</u> <u>Farmers'</u> Response <u>to</u>
     <u>Price</u>, Johns Hopkins Press, Baltimore, 1958
Neudecker,H., "Cournot's Dynamic Market Solution and Hosomatsu's Lemma:
     An Alternative Proof", <u>Review</u> <u>of</u> <u>Economic</u> <u>Studies</u>, 37 (1970)
Nikaido,H., "On the Classical Multilateral Exchange Problem", <u>Metro-</u>
     <u>economica</u>, 8 (1956)
Nikaido,H., <u>Convex</u> <u>Structures</u> <u>and</u> <u>Economic</u> <u>Theory</u>, Academic Press,
     New York, 1968
Okuguchi,K., "The Stability of the Cournot Oligopoly Solution: A
     Further Generalization", <u>Review</u> <u>of</u> <u>Economic</u> <u>Studies</u>, 31 (1964)
Okuguchi,K., "The Stability of Price Adjusting Oligopoly: The Effects
     of Adaptive Expectations", <u>Southern</u> <u>Economic</u> <u>Journal</u>, 35 (1968)
Okuguchi,K., "On the Stability of Price Adjusting Oligopoly Equilibrium
     under Product Differentiation", <u>Southern</u> <u>Economic</u> <u>Journal</u>, 35
     (1969)
Okuguchi,K., "On the Stability of the Cournot Oligopoly Equilibrium",
     <u>Economic</u> <u>Studies</u> <u>Quarterly</u>, 21 (1970a)
Okuguchi,K., "Adaptive Expectations in an Oligopoly Model", <u>Review</u> <u>of</u>
     <u>Economic</u> <u>Studies</u>, 36 (1970b)
Okuguchi, K., "A Comparison of Optimal Output in Risky Situations",
     <u>manuscript</u>, 1970
Okuguchi,K., "The Stability of the Stackelberg Duopoly Solution:
     Extensions of Kamerschen-Smith Results", <u>Economic</u> <u>Studies</u> <u>Quarterly</u>,
     22 (1971)
Okuguchi,K., <u>Kasen</u> <u>no</u> <u>Riron</u> (Theory of Oligopoly), Sobunsha, Tokyo,
     1971 (in Japanese)
Okuguchi,K., "Matrices with Dominant Diagonal Blocks and Economic Theory",
     <u>manuscript</u>, 1975
Okuguchi,K., "Further Note on Matrices with Quasi-Dominant Diagonals",
     <u>Economic</u> <u>Studies</u> <u>Quarterly</u>, 27 (1976)
Okuguchi,K., "Quasi-Competitiveness and Cournot Oligopoly", <u>Review</u> <u>of</u>
     <u>Economic</u> <u>Studies</u>, 40 (1973)

Olech,C., "On the Global Stability of an Autonomous System on the Plane", in Contributions to Differential Equations, 1, (1963)

Ostrowski,A.W., Über die Determinanten mit Überwiegender Haupudiagonale" Commentarii Mathematici Helvetici, 10 (1937)

Quandt,R.E., "On the Stability of Price Adjusting Oligopoly", Southern Economic Journal 33 (1967)

Rothschild,M. and J.E.Stiglitz,"Increasing Risk, I and II", Journal of Economic Theory, 2 and 3 (1970 and 1971)

Ruffin,R.J., "Cournot Oligopoly and Competitive Behavior", Review of Economic Studies, 38 (1971)

Samuelson,P., Foundations of Economic Analysis, Harvard University Press, Cambridge, 1947

Sandmo,A., "On the Theory of the Competitive Firm under Price Uncertainty", American Economic Review, 61 (1971)

Sato,R. and K.Nagatani, "The Stability of Oligopoly with Conjectural Variations", Review of Economic Studies, 34 (1967)

Shephard,W.G., "On Sales Maximizing and Oligopoly Behavior", Economica, 29 (1962)

Shubik,M., Strategy and Market Structure, John Wiley and Sons Ltd., New York, 1959

Simaan,M. and J.B.Cruz Jr., "On the Stackelberg Strategy in Nonzero-Sum Games", Journal of Optimization Theory and Applications, 11 (1973)

Stackelberg,H.von., Marktform und Gleichgewicht, Springer-Verlag, Berlin, 1934

Tarr,D., "Stability in a Cournot Market Characterized by Uncertainty", Western Economic Journal, 10 (1972)

Tarr,D., "The Effects of Adaptive Expectations on the Stability of Price-Adjusting Oligopoly", Southern Economic Journal, 41 (1975)

Telser,L., Competition, Collusion and Game Theory, Aldine, Chicago and New York, 1972

Theocharis,R., "On the Stability of the Cournot Solution of the Oligopoly Problem", Review of Economic Studies, 27 (1960)

Theocharis,R., "Some Dynamic Aspects of the Oligopoly Problem", Jahrbüchern für Nationalökonomie und Statistik, Band 178 (1965)

Uzawa,H., "The Stability of Dynamic Processes", Econometrica, 29 (1961)

Vol 59 J A Hanson, Growth in Open Economies V, 128 pages 1971

Vol. 60 H Hauptmann, Schatz- und Kontrolltheorie in stetigen dynamischen Wirtschaftsmodellen V, 104 Seiten 1971

Vol 61 K H F Meyer, Wartesysteme mit variabler Bearbeitungsrate VII, 314 Seiten 1971

Vol 62 W Krelle u G Gabisch unter Mitarbeit von J Burgermeister, Wachstumstheorie VII, 223 Seiten 1972

Vol 63 J Kohlas, Monte Carlo Simulation im Operations Research VI, 162 Seiten 1972

Vol 64 P Gessner u K Spremann, Optimierung in Funktionenraumen IV, 120 Seiten 1972

Vol 65 W Everling, Exercises in Computer Systems Analysis VIII, 184 pages 1972

Vol 66 F Bauer, P Garabedian and D Korn, Supercritical Wing Sections V, 211 pages 1972

Vol 67 I V Girsanov, Lectures on Mathematical Theory of Extremum Problems V, 136 pages 1972

Vol 68 J Loeckx, Computability and Decidability An Introduction for Students of Computer Science VI, 76 pages 1972

Vol 69 S Ashour, Sequencing Theory V, 133 pages 1972

Vol 70 J P Brown, The Economic Effects of Floods Investigations of a Stochastic Model of Rational Investment Behavior in the Face of Floods V, 87 pages 1972

Vol 71 R Henn und O Opitz, Konsum- und Produktionstheorie II V, 134 Seiten 1972

Vol 72 T P Bagchi and J G C Templeton, Numerical Methods in Markov Chains and Bulk Queues XI, 89 pages 1972

Vol 73 H Kiendl, Suboptimale Regler mit abschnittweise linearer Struktur VI, 146 Seiten 1972

Vol 74 F Pokropp, Aggregation von Produktionsfunktionen VI, 107 Seiten 1972

Vol 75 GI-Gesellschaft fur Informatik e V Bericht Nr 3 1 Fachtagung uber Programmiersprachen Munchen, 9 –11 Marz 1971 Herausgegeben im Auftrag der Gesellschaft fur Informatik von H Langmaack und M Paul VII, 280 Seiten 1972

Vol 76 G Fandel, Optimale Entscheidung bei mehrfacher Zielsetzung II, 121 Seiten 1972

Vol 77 A Auslender, Problèmes de Minimax via l'Analyse Convexe et les Inégalités Variationelles Théorie et Algorithmes VII, 132 pages 1972

Vol 78 GI-Gesellschaft fur Informatik e V 2 Jahrestagung, Karlsruhe, 2 –4 Oktober 1972 Herausgegeben im Auftrag der Gesellschaft fur Informatik von P Deussen XI, 576 Seiten 1973

Vol 79 A Berman, Cones, Matrices and Mathematical Programming V, 96 pages 1973

Vol 80 International Seminar on Trends in Mathematical Modelling, Venice, 13–18 December 1971 Edited by N Hawkes VI, 288 pages 1973

Vol 81 Advanced Course on Software Engineering Edited by F L Bauer XII, 545 pages 1973

Vol 82 R Saeks, Resolution Space, Operators and Systems X, 267 pages 1973

Vol 83 NTG/GI-Gesellschaft fur Informatik, Nachrichtentechnische Gesellschaft Fachtagung „Cognitive Verfahren und Systeme", Hamburg, 11 –13 April 1973 Herausgegeben im Auftrag der NTG/GI von Th Einsele, W Giloi und H -H Nagel VIII, 373 Seiten 1973

Vol 84 A V Balakrishnan, Stochastic Differential Systems I Filtering and Control A Function Space Approach V, 252 pages 1973

Vol 85 T Page, Economics of Involuntary Transfers A Unified Approach to Pollution and Congestion Externalities XI, 159 pages 1973

Vol 86: Symposium on the Theory of Scheduling and its Applications Edited by S E Elmaghraby VIII, 437 pages 1973

Vol 87 G F Newell, Approximate Stochastic Behavior of n-Server Service Systems with Large n VII, 118 pages 1973

Vol 88 H Steckhan, Guterstrome in Netzen VII, 134 Seiten 1973

Vol 89 J P Wallace and A Sherret, Estimation of Product Attributes and Their Importances V, 94 pages 1973

Vol 90 J -F Richard, Posterior and Predictive Densities for Simultaneous Equation Models VI, 226 pages 1973

Vol 91 Th Marschak and R Selten, General Equilibrium with Price-Making Firms XI, 246 pages 1974

Vol 92 E Dierker, Topological Methods in Walrasian Economics IV, 130 pages 1974

Vol 93 4th IFAC/IFIP International Conference on Digital Computer Applications to Process Control, Part I Zurich/Switzerland, March 19–22, 1974 Edited by M Mansour and W Schaufelberger XVIII, 544 pages 1974

Vol 94 4th IFAC/IFIP International Conference on Digital Computer Applications to Process Control, Part II Zurich/Switzerland, March 19–22, 1974 Edited by M Mansour and W Schaufelberger XVIII, 546 pages 1974

Vol 95 M Zeleny, Linear Multiobjective Programming X, 220 pages 1974

Vol 96 O Moeschlin, Zur Theorie von Neumannscher Wachstumsmodelle XI, 115 Seiten 1974

Vol 97 G Schmidt, Uber die Stabilitat des einfachen Bedienungskanals VII, 147 Seiten 1974

Vol 98 Mathematical Methods in Queueing Theory Proceedings 1973 Edited by A B Clarke VII, 374 pages 1974

Vol 99 Production Theory Edited by W Eichhorn, R Henn, O Opitz, and R W Shephard VIII, 386 pages 1974

Vol 100 B S Duran and P L Odell, Cluster Analysis A Survey VI, 137 pages 1974

Vol 101 W M Wonham, Linear Multivariable Control A Geometric Approach X, 344 pages 1974

Vol 102 Analyse Convexe et Ses Applications Comptes Rendus, Janvier 1974 Edited by J -P Aubin IV, 244 pages 1974

Vol 103 D E Boyce, A Farhi, R Weischedel, Optimal Subset Selection Multiple Regression, Interdependence and Optimal Network Algorithms XIII, 187 pages 1974

Vol 104 S Fujino, A Neo-Keynesian Theory of Inflation and Economic Growth V, 96 pages 1974

Vol 105 Optimal Control Theory and its Applications Part I Proceedings 1973 Edited by B J Kirby VI, 425 pages 1974

Vol 106 Optimal Control Theory and its Applications Part II Proceedings 1973 Edited by B J Kirby VI, 403 pages 1974

Vol 107 Control Theory, Numerical Methods and Computer Systems Modeling International Symposium, Rocquencourt, June 17–21, 1974 Edited by A Bensoussan and J L Lions VIII, 757 pages 1975

Vol 108 F Bauer et al , Supercritical Wing Sections II A Handbook V, 296 pages 1975

Vol 109 R von Randow, Introduction to the Theory of Matroids IX, 102 pages 1975

Vol 110 C Striebel, Optimal Control of Discrete Time Stochastic Systems III 208 pages 1975

Vol 111 Variable Structure Systems with Application to Economics and Biology Proceedings 1974 Edited by A Ruberti and R R Mohler VI, 321 pages 1975

Vol 112 J Wilhlem, Objectives and Multi-Objective Decision Making Under Uncertainty IV, 111 pages 1975

Vol 113 G A Aschinger, Stabilitatsaussagen uber Klassen von Matrizen mit verschwindenden Zeilensummen V, 102 Seiten 1975

Vol 114 G Uebe, Produktionstheorie XVII, 301 Seiten 1976

Vol. 115· Anderson et al, Foundations of System Theory Finitary and Infinitary Conditions. VII, 93 pages. 1976

Vol 116. K. Miyazawa, Input-Output Analysis and the Structure of Income Distribution IX, 135 pages 1976

Vol 117· Optimization and Operations Research Proceedings 1975 Edited by W Oettli and K Ritter IV, 316 pages 1976

Vol 118· Traffic Equilibrium Methods, Proceedings 1974 Edited by M. A. Florian XXIII, 432 pages. 1976

Vol 119 Inflation in Small Countries Proceedings 1974 Edited by H Frisch VI, 356 pages 1976

Vol 120· G Hasenkamp, Specification and Estimation of Multiple-Output Production Functions VII, 151 pages 1976

Vol 121. J W Cohen, On Regenerative Processes in Queueing Theory IX, 93 pages 1976

Vol 122 M S Bazaraa, and C M Shetty,Foundations of Optimization VI 193 pages 1976

Vol 123 Multiple Criteria Decision Making Kyoto 1975 Edited by M Zeleny XXVII, 345 pages 1976

Vol 124 M J Todd The Computation of Fixed Points and Applications VII, 129 pages 1976

Vol 125· Karl C Mosler Optimale Transportnetze Zur Bestimmung ihres kostengunstigsten Standorts bei gegebener Nachfrage VI, 142 Seiten 1976

Vol 126. Energy, Regional Science and Public Policy Energy and Environment I Proceedings 1975. Edited by M Chatterji and P Van Rompuy VIII, 316 pages 1976

Vol 127· Environment, Regional Science and Interregional Modeling Energy and Environment II Proceedings 1975 Edited by M Chatterji and P Van Rompuy IX, 211 pages 1976

Vol 128 Integer Programming and Related Areas A Classified Bibliography Edited by C Kastning XII, 495 pages 1976

Vol 129 H-J Luthi, Komplementaritats- und Fixpunktalgorithmen in der mathematischen Programmierung Spieltheorie und Okonomie VII, 145 Seiten 1976

Vol 130 Multiple Criteria Decision Making, Jouy-en-Josas, France Proceedings 1975 Edited by H Thiriez and S Zionts VI, 409 pages 1976

Vol 131 Mathematical Systems Theory Proceedings 1975 Edited by G Marchesini and S K Mitter X, 408 pages 1976

Vol. 132· U H Funke, Mathematical Models in Marketing A Collection of Abstracts XX, 514 pages 1976

Vol. 133 Warsaw Fall Seminars in Mathematical Economics 1975 Edited by M W Loś, J Loś, and A Wieczorek V 159 pages 1976

Vol 134 Computing Methods in Applied Sciences and Engineering Proceedings 1975 VIII, 390 pages 1976

Vol 135 H Haga, A Disequilibrium – Equilibrium Model with Money and Bonds A Keynesian – Walrasian Synthesis VI, 119 pages 1976

Vol 136 E Kofler und G Menges, Entscheidungen bei unvollstandiger Information XII, 357 Seiten 1976

Vol 137 R Wets, Grundlagen Konvexer Optimierung VI, 146 Seiten 1976

Vol 138 K. Okuguchi, Expectations and Stability in Oligopoly Models VI, 103 pages 1976